A Handbook of
Writing for Engineers

To Wolfgang
Marta and Richard
Abigael, Georgia and Stefan

A Handbook of Writing for Engineers

Joan van Emden

Second Edition

MACMILLAN

© Joan van Emden 1990, 1998

First edition 1990
Reprinted three times
Second edition 1998

Published by
MACMILLAN PRESS LTD
Houndmills, Basingstoke, Hampshire RG21 6XS
and London
Companies and representatives
throughout the world

ISBN 0–333–72807–6 paperback

A catalogue record for this book is available
from the British Library.

This book is printed on paper suitable for recycling and
made from fully managed and sustained forest sources.

10 9 8 7 6 5 4 3 2 1
07 06 05 04 03 02 01 00 99 98

Printed in Malaysia

Contents

Preface to Second Edition

Writing is probably that part of a working life which engineers most dislike. Unfortunately, the more successful they are in their profession, the more time they will spend sitting at the wordprocessor preparing to convey technical information to colleagues, customers and clients.

Modern technology makes the chore of writing easier, but does not help engineers who want to present convincing arguments clearly and concisely. This book looks at the basic skills of spelling, punctuation and grammar, and also at appropriate style, conventional formats and the effective presentation of written information. Its aim is to give confidence to a wide range of engineers, undergraduates and research students, managers and consultants – anyone who in the course of a working life needs help and encouragement in facing the challenge of writing well.

In this second edition, as in the first, many of the examples given are 'real-life', taken from university work or from company documentation. I continue to be grateful to all the hundreds of anonymous engineers whom I meet on technical writing courses, who have generously given me permission to use their writing and explained it to me. In particular, I should like to thank the Engineering Department of the University of Reading, especially Dr J. B. Grimbleby and Dr A. J. Pretlove, for their help and support; similarly, I am indebted to Dr Martin Coutie and Dr Martyn Ramsden for permission to use some of their writing. Any errors which remain are my own.

JOAN VAN EMDEN

Reading
1997

1 Introduction

Reading ■ discussing ■ writing ■ confidence ■ reader goodwill: identifying readers and objectives ■ getting started

> *Reading maketh a full man; conference a ready man; and writing an exact man.*
>
> Sir Francis Bacon, *Essays* (1597)

Communication as described by Sir Francis Bacon in the late sixteenth century is much the same as the communication of a practising engineer – man or woman – today. Reading, discussing and writing take up a large part of a working life, and through these activities knowledge is broadened, abilities are sharpened, reactions become more focused; the experience and expertise of the professional engineer are presented precisely and effectively.

☐ Reading, discussing and writing are essential aspects of communication for the practising engineer

This book is primarily concerned with the third aspect, writing, but the other two are equally important. Engineers must find out what is happening in their field, nationally and internationally; they must keep up to date with current practice and study the exact requirements of their companies and their clients. They must read the relevant documentation and be ready to respond to it if their knowledge is to be 'full', that is, sufficient to allow them to make appropriate decisions.

It would be perhaps brave and certainly foolhardy to take all such decisions in isolation. Bacon's second requirement, 'conference', involves engineers in meeting people, in talking to their clients, giving instructions and making presentations, and in discussing day-to-day problems with other engineers. Co-operation and mutual support result from such interaction, especially if 'conference' includes the underrated ability to listen well.

Writing, says Bacon, makes the writer 'exact'. In transforming ideas into written words, engineers have to make choices. They have to analyse exactly what is to be expressed, to identify the readership, and to decide on the appropriate format and style. In the education and training which made them engineers, they discovered the need for careful, precise and logical thought. In learning to be writers, they must apply the same criteria. Technical knowledge has to be communicated accurately, and as engineers write letters, outline their proposals, prepare specifications or plan reports, they must again be careful, precise and logical. They must have constantly in mind the needs of their readers, adjusting the amount of detail in the light of their readers' needs and presenting the material in a logical form which can be identified easily and used with confidence.

Confidence is a key word in this book. Engineers, generally speaking, prefer to do the job rather than to write about it. When faced with the blank computer screen, they may experience panic. Which words should they choose and in what order? What are the conventions they should follow? How can they hold the reader's attention? How can they write convincingly?

☐ Care, precision and logic are necessary to thinking and to writing

Writing for Engineers provides guidance in the use of words, construction of sentences and organisation of paragraphs. It also looks at some of the most important types of format, both traditional such as the letter and more modern such as the fax, and discusses the conventions which should be followed. Most of all, this book aims at giving engineering writers the confidence that they are conveying their information accurately and in a readable style. They are thus not only able to do the work, but also to write about it in a professional way.

Nevertheless, accuracy by itself is not enough to hold the attention of readers and to convince them of the writer's point of view. Engineers have been known to go to extreme lengths in order to make an impact. One young professional, faced with the problem of presenting monthly reports which seemed to be regularly ignored, wrote his report in verse. His manager was, not surprisingly, surprised. He called the young man into his office and held forth at length about this aberrant behaviour. When he paused to ask what the writer had to say for himself, he was even more surprised and, one hopes, abashed at the reply: 'This is the first time that you've taken the trouble to discuss my report with me.'

Shock tactics apparently work, but are not recommended. If reports are written concisely, if the information is easily assimilated and the format well chosen, they will probably be read. What is certain is that if they are longwinded and unstructured, they will be ignored for as long as possible and finally read unhappily, if at all.

☐ Good writing generates reader goodwill.

The best advice for the prospective writer is, then, as follows: identify your readers, know what they already know and what they need to know, find out how much technical knowledge they are likely to have and what their involvement with the project is; have full and accurate information at your disposal; formulate your objectives (what you want to get out of this piece of writing); have confidence in yourself, and then write.

☐ Identify your readers and their objectives, and your own objectives, before beginning to write.

A last word: don't begin at the beginning. The first sentence or paragraph is almost always the most difficult. Choose a simple, straightforward factual section which you feel comfortable with, and write it first. Your confidence will receive a boost, and by the time you reach the first section of your document, you will have had considerable practice in the art of good writing. You are already 'exact'.

☐ Don't begin at the beginning.
☐ Confidence grows with the practice of good writing.

Key ideas

■ Reading, discussing and writing are essential aspects of communication for the practising engineer (p. 1).

■ Care, precision and logic are necessary to thinking and to writing (p. 2).

■ Good writing generates reader goodwill (p. 3).

■ Identify your readers and their objectives, and your own objectives, before beginning to write (p. 3).

■ Don't begin at the beginning (p. 4).

■ Confidence grows with the practice of good writing (p. 4).

2 Vocabulary

Choice of words ■ an international language ■
American English ■ prefixes ■ accuracy ■ pairs of
words ■ synonyms ■ precision ■ spelling new words ■
simple language ■ avoiding clichés and slang ■ jargon

Some people like words and others like numbers. The sad
truth is that those who like words rarely also like numbers,
and *vice versa*. Engineers often have love affairs with figures
(having passed a range of mathematically orientated exam-
inations in order to become engineers) but tend to feel that
words are out to get them. As a result, they are uneasy about
writing continuous prose, and long for the release of a
friendly equation.

Words are indeed difficult in the English language. They
are spelt in odd ways, are often pronounced differently from
the way they look, may sound the same as each other but
have different meanings, and there are so many of them. It is
a great advantage to have a language which is rich in synonyms (words which mean more or less the same thing), but
there is a catch in this: which words should we choose?

The answer to this question is in four parts. We should
choose words which are understood by the reader, familiar
without being hackneyed, accurate, and as simple as the
subject matter allows. The first of these points is sometimes
overlooked. Engineers often write for a wide audience, not all
of whom will have the same specialist knowledge. One of the
early, preparatory questions to be asked before beginning to
write any document is about the knowledge and experience
levels of the intended readership. Will most of the readers

understand the technical terms used? There is of course a danger in over-explaining, seeming to patronise the reader, but the more common problem is bewildering those who have different expertise from that of the writer.

The best compromise in most cases is to include a glossary of technical terms and abbreviations which might be unknown to at least some readers. Such glossaries are common in reports and specifications, and are helpful to those who need them, while, for those who have the appropriate specialist knowledge, remaining unobtrusive. In the case of a shorter document such as a letter or memo, it is usually better simply to write the terms in full or to include a short explanatory phrase rather than to risk causing irritation to the recipient.

However, the readership for some technical writing will be much wider; an engineer may be writing for an international readership. On the whole, the technical terms will cause less difficulty than the 'simple' words in between – many engineers receive their technical training in English. It is words or expressions which are used by native English speakers in a colloquial way, or which refer to concepts local to western European culture which cause the greatest difficulty.

An article in the *Guardian* about the ITN World News[1] points to some of the problem areas. International news broadcasts refer to the Chancellor of the Exchequer as the 'Finance Minister', which is clear to listeners who have Finance Ministers in their own countries but who would not readily recognise the Chancellor in his usual British guise. Expressions such as a 'wave of strikes' or 'people turned out' are rejected; a term such as 'Bank Holiday' will be transformed to 'national holiday'. There is never a play on words on the World News.

The same principles apply to engineering writing for a readership which is not primarily English speaking. Expressions which are 'local' should not be used, and the writer should show sensitivity in avoiding, as far as possible, expressions which are unnecessarily complicated, for example 'on no account do this' can be expressed more simply and clearly as 'do not do this.' Nevertheless, if the only accurate word or expression is a rare or complex one, it must be used, as accuracy cannot be sacrificed to familiarity. In such cases, a note might be appropriate, or, better still, a diagram. Illustrations are on the whole more easily and widely understood than paragraphs of explanation. Writers should always be aware of their readers, remembering that engineering information is sometimes translated into other languages; there is no point in making life harder than necessary for the translators.

☐ Use words and expressions which readers understand; if appropriate, include a glossary.

Many engineers work for American or for multinational companies, and it is worth remembering that we often fail to share our common language. Differences in spelling are well known: *color, traveling, center,* are all acceptable in American English – and may as a result be the forms desired by the computer spellcheck. When meanings are different, the problem is more serious. A *UK gallon* is about a sixth more than a *US gallon,* an English *chemist* is an American *druggist,* although in both countries the word *chemist* is used for a scientist whose specialism is chemistry. A billion is now standardised at 1000 million on both sides of the Atlantic, but the British *ground floor* (American *first floor*) may still cause misunderstanding, and *to table a motion* at an American meeting is to set it aside – exactly the opposite to the British meaning. The only guideline is that if you find yourself arguing with an American colleague, check that it is not the language which is dividing you rather than the subject of debate. If you frequently exchange written material, keep an American dictionary next to the English one.

English itself has assimilated a great number of foreign words and phrases; it also uses some which have not been assimilated but remain 'foreign' and are often put into italics as a result. This book uses *vice versa* in that way, for instance. In the past, professional writing was full of such expressions, especially in letters (the *5th ult., viz.*) and in published references (*op. cit., ibid.*). Many people nowadays do not understand such expressions, which are seen as pompous, 'showing off'. Unless there is a good reason for using a foreign expression, or it is widely accepted, it is better avoided. However, prefixes (additions to the beginnings of words) often have their origins in foreign languages, and it is useful to be able to recognise them. *Bi* means two or twice (originally Latin), and knowing that, we can more easily understand words like *bilateral, bipartite* or *binary*. Other useful prefixes are:

ambi = both	(ambidextrous, ambivalent)
ante = before	(antecedent, anteroom)
anti = against	(anticlockwise, antisocial)
auto = self	(automatic, autobiography)
bene = good, well	(benevolent, benefit)
circum = around	(circumference, circumscribe)
con = with	(concentric, concord)
contra = against	(contradict, contravene)
extra = beyond	(extraordinary, extramural)
dia = through	(diameter, diagram)
homo = the same	(homogeneous, homologise)
hyper = beyond, above	(hyperactive, hyperbole)
hypo = under, below	(hypothesis, hypocrisy)
inter = between	(interdependent, intervene)
intro = inside	(introvert, introduction)
mal = bad	(malfunction, malevolent)
meta = change	(metamorphic, metabolism)
para = beside	(parallel, parameter)
poly = many	(polygon, polymer)
retro = backwards	(retrograde, retroact)
ultra = beyond	(ultrasonic, ultraviolet)

The reader must understand the words that are chosen, and they must be used accurately. Accuracy has two aspects: words must be used so that the intended meaning is conveyed precisely; the words which appear in the text must be the exact words which the author intended. Both criteria present problems.

Inaccurate use of words is usually the result of imprecise thinking. The engineer who produced the following example did not consider exactly what needed to be said before putting pen to paper:

> *The purpose of this memo is to notify all company*
> *personnel involved with the Zero project of the present*
> *situation regarding outstanding items not delivered to the*
> *customer yet for one reason or another.*

As a result of this failure to identify the message, the writer uses unnecessary and confusing words. 'The purpose of this memo is to notify' is self-evident. Items which have not been delivered to the customer yet *are* outstanding items. The sentence peters out with the strange comment 'for one reason or another'. This sounds worrying – all sorts of things are going wrong – but it means nothing. Words are used so loosely that the recipients must have felt as if they were holding tightly onto jelly, which oozed through their fingers.

The writer needs to clarify the message first of all:

> *We've problems of late delivery on the Zero project.*

The format for this message is chosen: a memo. This starts with the conventional headings (see p. 70) which are probably printed on the page:

To *Zero Project personnel*   **Date** *10 November 1996*
From *Jim Twigg, Project*   **Subject** *late deliveries*
         Manager

All the writer now has to do is to give details of the problems and, if possible, suggested solutions. The original sentence has disappeared completely, to be replaced by concise, accurate writing. The recipients are put in the picture at once.

Jim Twigg might have chosen to send his message by e-mail, and the same principles would have applied. The medium is swift and informal, but the need for precision and economy of words is still paramount – too often the very ease of e-mail persuades its users to become wordy or to ramble away from the message (see p. 68). Time is important for the recipient as well as for the sender.

Illogicality in writing can result from similar poor planning. Words convey the wrong meaning, for example:

> *As the metal becomes harder and hence an increase in carbon content, the metal tends not to increase its reduction, but instead the area is less than a metal with less carbon.*

The underlying mistake in this sentence is perhaps the writer's ignorance of the measurement 'reduction of area', that is, the ductility of the metal. While there has been an attempt to relate hardness to carbon content, the message has not been thought through. The hardness of the metal does not, as is suggested, result ('hence') in an increase in carbon content; rather, as the carbon content increases, we find an increase in hardness. The word 'hence' is totally misleading. The oddly expressed 'tends not to increase its reduction' is confusing, not least because of the juxtaposition of 'not', 'increase' and 'reduction'. The last part of the sentence is probably an attempt to reiterate the first part, but the writer is too ill at ease with the information to be able to say anything on the subject clearly. The sentence might be written simply as:

> *As the carbon content increases, so does the hardness, and at the same time the ductility of the metal is reduced.*

☐ Identify the message and plan its expression.

English contains many words which sound alike but which have different spellings and different meanings. A writer in a hurry can easily confuse words which under other circumstances would easily be distinguished. Such words are *stationery* (writing paper) and *stationary* (without movement), *principal* (chief) and *principle* (underlying rule or moral basis), *draft* (rough version) and *draught* (current of air). Sometimes a group of three words can cause confusion, such as *cite* (quote), *sight* (vision) and *site* (area of land). The spellcheck is no help in drawing attention to a misuse of these words, and perhaps as a result, such mistakes are surprisingly common. The result can be faintly comic, as in the case of the student who wrote:

Diamond is tough and is covalent, but other chain polymers may be extremely week.

Some pairs of words sound almost, but not quite, the same and are easily confused, such as *moral* (ethical) and *morale* (emotional condition) or *personal* (belonging to the individual) and *personnel* (staff). Some such difficulties arise regularly in engineering writing, for instance:

accede and *exceed*:	*accede* = assent to ('he acceded to my request') *exceed* = surpass ('he exceeded the speed limit')
access and *assess*:	*access* = entry ('access to the building') *assess* = weigh up ('assess the capability of the machine')
principle and *principal*:	*principle* = moral principle ('his principles prevented his cheating') *principal* = main, chief ('the principal reason for the change...')

A further common example is slightly more complicated: *affect* and *effect*. The difficulty arises because *effect* can be both noun and verb, and also because many people do not pronounce the words clearly, probably because they are not sure which is which!

Affect is a verb, which means 'to have an influence on', as in:

Studying engineering to an advanced level has affected his job prospects.

Effect can be a verb, meaning 'to bring about', as in:

> *Study, hard work and experience combined to effect an improvement in his career prospects.*

Effect can also be a noun, meaning 'an influence' or 'the result', for example:

> *His overseas experience has had an effect on his career pattern.*
> *The effect of his hard work has been rapid promotion.*

Other pairs of words which often give trouble are those which depend for their spelling on the way in which they are used. Some have a useful difference in pronunciation, such as *advice/advise* and *device/devise*, while others sound the same, such as *practice/practise* and *licence/license*. In each case, the noun has a *c* and the verb has an *s*. For example:

> *I can advise you to study engineering, but will you take my advice?*
> *He perfected a device for sounding the alarm but could not devise a way of ensuring that people would respond.*

Sometimes words with a similar derivation have nowadays moved further apart:

> *disinterested* = impartial, without prejudice
> *uninterested* = having no interest in (almost bored)
> *complex* = involved, technically difficult
> *complicated* = mixed up, difficult to untangle

Often associated with these words are *imply* and *infer*.

> *imply* = suggest, hint
> *infer* = understand, assume a meaning

This pair could be thought of as the opposite ends of a conversation:

> *I implied that she might soon be promoted, and she inferred that a senior position was becoming vacant.*

In all these cases, the problems are obvious but the solutions much less so. A writer of technical information must be aware of words and of the ways in which they can be confused, alert to the look and sound of words, careful and attentive in both listening and writing. Pronunciation can have a direct impact on spelling: a mis-spelling such as '*failiure*' is the result of saying the extra 'i' after the 'l' and then hearing and writing it; '*presence*' and '*presents*' can sound exactly the same if they are carelessly spoken; the common mis-spelling '*maintainance*' is usually the effect of pronouncing the word that way rather than, correctly, '*maintenance*'. As with most problems of English, critical reading will help, together with a determination to make friends with words as well as with numbers.

□ Choose words with care; read critically.

Earlier in this chapter, the comment was made that English is particularly rich in synonyms – words which have exactly the same meaning, or sometimes a very slight difference in meaning. However, apart from the dictionary meaning, there is also the feel of a word; we know almost instinctively which word to choose for a particular context. *Difficult, hard* and *troublesome* can all mean the same thing, and yet we would not be likely to describe a difficult question as troublesome or a troublesome toothache as difficult.

Some synonyms cause particular problems because, while their meanings are the same, the implications are different. A *request* is different from a *desire to know* or a *demand to know*; the speaker might prefer to be thought of as *forceful*, while listeners might use the word *aggressive*. It depends on the point of view! We choose words not only for their meaning but also for their implications: so we may congratulate a friend for showing *determination* while deprecating the *stubbornness* of an opponent. We may hesitate to *recommend* a course of action, and so we *suggest* it, which sounds milder – and of course if things go wrong, we can always claim that it was 'only a suggestion'.

Dictionaries can be dangerous. They are rarely capable of showing such fine shades of meaning, and we have to be sensitive to the impact of the words we choose. Words also change their meaning according to their context. If we look up *power* in the dictionary, we find a wide range of meanings, including *ability to act, vigour, energy, government, personal ascendancy, authorisation, influential person, magnifying capacity,* and so on. The writer must be sure that the intended meaning is clear to the reader. We know what we mean if we talk about 'six to the power of ten' or 'switching on the power supply' or 'the power of the pianist's interpretation of a concerto', but if we say that we have the *power* to make a particular piece of information known, do we mean the *authority* or just the *ability*? Our reader might not be sure which is the correct interpretation, and we must clarify our intention.

Casual speech sometimes misuses words in a way which is not acceptable in the written language. *Aggravate* means 'to make worse' (not 'to annoy'). *Unique* means that there is only one example in existence, not that it is rare. *Former* and *latter* are the first and second of two, never of more than two; the correct use is shown in the following example:

> *The lights and the steering need to be adjusted: the former can be done at once but the latter will take us a bit longer and so we will need the car tomorrow as well as today.*

Other groups of words are often misused: *number, majority* and *fewer* always refer to several objects or people, while *amount, greater* (or *lesser*) *part* and *less* refer to only one object or person. For example:

> *There was a large number of delegates, and the majority favoured the new agreement. Fewer than a hundred objected.*

> *There was a large amount of work still to be done, the greater part of which, on this occasion, had to be completed during the night. Fewer workers and less time will be needed in future.*

Accuracy of expression also means that the words chosen are in themselves precise. Words and phrases such as *fairly, quite, rather, to a limited extent* and *in due course* produce a vague, hesitant impression, while *very, extremely, mainly* and *substantially* sound important but convey no precise picture. For example:

> *Our experiments were fairly successful and we are generally hopeful that we shall be able to make the results public in due course.*

This means very little. What is 'fairly successful?' Some of the experiments, or on some occasions? What percentages? 'Generally hopeful?' All of them or only most of them? Most of the time? When they aren't feeling depressed? 'In due course?' Next week? Next year? Eventually, if we're lucky?

While it would be wrong to suggest that such words and phrases are always unhelpful and should therefore never be used, they can be irritating to the reader who is looking for precise information. If it is possible to be exact, then be exact:

> *We were successful with 60% of our experiments, and provided that there are no unforeseen problems, we shall make the results public within the next two months.*

So now we know!

Engineering information must always be accurate; it must also be conveyed accurately. Lack of precision in, for example, a specification or a report, can have serious implications, and it is often the case that while the technical terms are correctly used, the small English words which appear in between the technical words cause ambiguity or misunderstanding. Two short examples illustrate this:

> *In future, the company will need less skilled workers for the night shift.*

> *In future, the company will need fewer skilled workers for the night shift.*

Which meaning was intended? In the first sentence, the same number of workers could have less training or experience and so less skill; in the second, the company will not need so many workers, perhaps because they are highly skilled. There is clearly a significant difference.

A similar problem can arise, again because of imprecise use of words, in the following example:

Maintenance shall be carried out at regular intervals or where there is evidence of malfunction.

Maintenance shall be carried out at six monthly intervals and whenever there is evidence of malfunction.

'At regular intervals' means little more than 'in due course'; how regular is regular? Every ten years? 'Or' does not really signify an alternative, but an additional reason for maintenance. 'Where' is not used in its usual meaning of 'the place where'; in casual conversation, we often confuse 'where' and 'when' and little harm is done; in the above sentence, the difference between place and time is crucial. As we can see from these examples, words which seem in themselves to be unimportant can change the meaning for the reader.

Words must not only be used accurately, they must also appear accurately in the text. This means that they are spelt correctly and that the wordprocessed version says exactly what the writer intended. Checking, an important stage in the production of any document, is discussed later (see p. 119). It is enough at this stage to point out that unchecked work can look very odd on the page, as in the case of the engineer who wrote, in awful handwriting, *50ft*, and was appalled to discover that it had been reproduced (unchecked) as *soft*.

Spelling is a nightmare to many engineers, slightly mitigated nowadays by the computer spellcheck. As we have seen, this has great advantages but also limitations, and it is still worth having a dictionary on the desk. Some words will in any case escape because they are too new or too technical for either aid. It is worth adding such words, if they are to be used frequently, to the computer dictionary – but with the reservation that the spelling should be checked first. It is not unknown for words to be added incorrectly, causing continuing problems.

Both management and modern technology produce new words or reuse old words in a new form. Some have limited application, while others become widely known and used. For example, most people in the past would have registered *mouse* as a small furry animal; many people now would just as quickly think of it as a device for selecting objects or defining positions on the screen. Sometimes modern developments give rise to complicated- sounding expressions, such as *negative corporate worth*, defined as the disadvantage for a subsidiary company of belonging to a larger corporation. Technology seems to favour composite words made up of words which in themselves are short and simple: *workstation, spreadsheet* are two examples. On the whole, such words are joined up without a hyphen, although they can become just as top-heavy as too many words containing hyphens. It is not easy to read a sentence like the following:

This is network-transparent, operating-system-independent and portable ... it promises to revolutionise high-end computing.

If the writer is in doubt about how to write new words, the best guidance is probably to check what the major journals in the field are doing. The Engineering Institutions are likely to face such questions very early on, and their decisions may well set the standard for future use.

Since English is a living language, usage that was once 'wrong' may gradually have become acceptable. The most common example of this is *data*, a plural word (singular *datum*) nowadays generally used as a singular – the rationale for this being that data is a collective noun like group or committee. However, such developments, at least in the short term, do not influence similar words, and *strata, criteria* are still considered to be plural (singulars *stratum, criterion*).

Old words can be just as worrying as new words, and there are many common words in English which are just difficult to spell. There is little logic about English spelling, and the rules that do exist generally have a great many exceptions. The only one which is worth remembering is 'i before e except after c, as long as the sound is ee'; this rule (provided it is remembered in its entirety) is actually helpful, as there are many instances in which it is followed (*believe, receive, height*) and comparatively few exceptions (*seize*). It is always worth making up mnemonics for one's personal list of most hated words; the more silly the idea, the easier it is to remember it (for instance, *liaison* needs more than one person and so contains two i's).

☐ Use the spellcheck and a dictionary, both with great care!

Words should be as simple as the context allows. Some writers seem to aim at confusion, as, for instance, does the author of the following guideline for the numbering of tables in reports:

Tables will be given successive decimal integer numbers of ascending value starting at unity.

This means no more than:

Number tables sequentially from 1.

Indeed, 'from 1' would almost certainly be assumed by the reader, and could be left out. Fourteen words have been reduced to three!

Yet the (real-life, not made up) sentence quoted above in its original form is typical of much engineering writing. Instead of making their point in a simple, straightforward way, writers wrap up their message in long sentences made up of words which are intended to sound impressive. It is amazing what peculiarities can result. An American was recently described as being *tasked with a mission*; the most interesting feature of this expression is that noun and verb could be exchanged – he could have been *missioned with a task*. He was, in fact, given a job.

There seem to be various reasons for such pompous language. One is undoubtedly the need to hide unpleasant or embarrassing information, as in the following example:

Following our recent meeting, we feel we must put in writing which we believe to be the justifying factors leading to our proposed modest increases in costs.

Clearly, the writer is uncomfortable. Whatever should have been put in writing has actually been left out ('the factors' or some such expression), and the unfortunate word 'modest' has been dragged in to make the writer feel better. Modest by whose standards? Why should the company concerned stress that the rise was 'justified'? What have the writer's feelings and beliefs got to do with the situation? The use of 'recent' suggests the sub- text 'I should have done this sooner and am embarrassed to give the date'. It would have been better to come straight to the point:

In the light of our meeting on 21 January, I confirm our need to raise our prices.

If it is really necessary to say why this has happened, a short sentence of explanation could follow.

Pompous words are also used because they are felt to be more influential than everyday words. Writers *acquaint [their] readers with* or *advise [their] readers of,* rather than simply telling or informing them. They *render assistance* rather than *give help* and they *subject to examination* rather than *examine.* They *are in a position to undertake* (*they can* or *are able to*), and they *peruse* documents when most people would read them; they prefer to *utilise* rather than to *use.* In so doing, they waste everybody's time and irritate the reader; they do not impress. Such inflated language, full of self-importance, may disguise the information. The engineer who wrote

The number of samples tested is important. If it is too small, poor results can be concealed or indicate erroneous behavioural characteristics.

failed to analyse what was written. If the sample is too small, results give false indications, that is, give poor results. The sentence goes in a circle. The meaning behind all this was probably the obvious fact that

> *A sufficient number of samples must be tested before the result can be considered valid.*

☐ Use simple words as far as the context permits.

The final criterion in the choice of words is that they should be familiar to the reader without being hackneyed. One of the problems of the pompous words discussed above is that they are also rare and therefore not easily understood. We have been brought up with a certain amount of 'official-speak' from public organisations, such as *tender the correct fare* or *passengers alight here* which are recognisable, but not natural. Few of us would use *tender* or *alight* in such a context. Writers should use words which readers can recognise and feel at home with.

However, it is possible to go too far the other way. Expressions which are over-used become clichés, dull and lifeless phrases which have ceased to make any impact. *At the end of the day* meaning *finally* or *eventually*, *at this present moment in time,* meaning *now, leave no stone unturned* meaning *try hard*, are all such expressions, used to the point at which they distract or irritate the reader and so undermine confidence in the information itself.

Slang belongs only to the spoken word and then only in casual conversation between friends. In writing, it jars on the reader and interrupts concentration. There is an example of this effect on p. 47, in the passage beginning 'It should be noted...'. The writer used the (in this context) pompous word 'terminate' and then the slang expression 'start again from scratch'. The transition is unpleasant, as the reader is jerked from over-formal language into everyday speech. Slang has no place in engineering writing.

☐ Clichés and slang distract the reader and should be avoided.

Jargon, however, comes in two kinds, one acceptable and the other not. The former is professional, understood by other engineers in the same discipline, but not part of the everyday language of the non-specialist. It would be impossible to manage without this sort of jargon, as it allows an expert to communicate easily with other experts. Both writers and readers take a range of such expressions for granted, assuming mutual understanding. Examples of such jargon in the field of expert systems are *antecedent, backward chaining, blind search, control structure* and *declarative knowledge representation.* Two possible dangers arise from the use of such terms. One is that the reader will not understand, and the engineering writer must always be alert to the need for acceptable communication with the 'outside world', for instance with clients who have less specialist knowledge than the writer.

The other danger is more insidious: both writer and reader may assume understanding when none exists. For example, in the list given above, *antecedent* means the left-hand side of a production rule, that is, the pattern needed to make the rule applicable. The dictionary definition of antecedent is 'preceding circumstance' or, in logic, 'the part of a conditional proposition on which the other depends'. We can see a relationship between the expert system definition and the lay definition, but they are not interchangeable. Expert jargon must be used with care.

The other kind of jargon is widespread and horrible. It uses words unnecessarily, sounds pompous and is either without meaning or conceals meaning. There are temporarily popular words and phrases such as *function* ('the engineering function' is the job of the engineer), *facility* (as in 'the manufacturing facility' meaning the factory), and *situation,* which may be added to almost anything – a crisis situation, a late delivery situation, an overmanning situation. It doesn't mean anything. Such terms can be used effectively to disguise meaning, as in:

> *The interactive function of the project manager with the team was such that it necessitated urgent implementation of the staff transfer policy.*

In other words, the project manager could not get on with the team and someone had to be moved, fast. There are even buzz-word jargon generators (a nice piece of jargon in itself) which allow expressions to be put together in any order so that they form impressive but meaningless sentences:

> *A constant flow of effective information will maximise the probability of project success and minimise the cost and time required for standardisation of anticipated documentation.*

If this means anything, it means 'let's get the communication right!'

☐ Professional jargon must be shared with the reader.
☐ 'Popular' jargon should be avoided.

Key Ideas

■ Use words and expressions which readers understand; if appropriate, include a glossary (p. 7).

■ Identify the message and plan its expression (p. 10).

■ Choose words with care; read critically (p. 13).

■ Use the spellcheck and a dictionary, both with great care! (p. 18).

■ Use simple words as far as the context permits (p. 20).

■ Clichés and slang distract the reader and should be avoided (p. 20).

■ Professional jargon must be shared with the reader (p. 22).

■ 'Popular' jargon should be avoided (p. 22).

---

3 Sentences and Punctuation

Sentences: Definition of a sentence ■ sentence length ■ simple and compound sentences ■ sentence structure ■ ordering the information ■ unrelated participles ■ incomplete verbs ■ split infinitives ■ confused constructions ■ redundant words and phrases ■ singular and plural ■ some tricky expressions ■ negative writing

Punctuation: full stops □ exclamation marks □ question marks □ semicolons □ colons □ commas □ quotation marks □ dashes □ brackets □ hyphens □ apostrophes

Sentences

A sentence is as long as a piece of Sellotape, and, for some people, more sticky. This is a useful analogy, as it avoids the cliché ('as long as a piece of string') and at the same time gives positively helpful information. Sellotape is an excellent product, provided that *the length of tape is right for its purpose*. Too short a piece, and the paper fails to fasten the parcel; too long a piece, and tape, paper and fingers stick together in a nasty mess.

Sentences are much the same. A very short sentence may fail to give the required information, or it may leave out essential elements such as a verb. A very long sentence often confuses itself, its writer and its reader, and ends up as a nasty mess. Sentences must be the right length for their purpose.

Before looking more closely at sentence length and construction, we should say what a sentence is.

Definition of a Sentence

☐ A sentence is a group of words which makes sense in itself.

☐ A sentence contains at least one main item of information to which various subsidiary ideas may be attached.

☐ A sentence must contain at least one complete verb.

The first aspect of this definition is most important, and many mistakes would be avoided if the writer asked, 'Does this make sense?' If it does not, it is not a sentence. Complete understanding may depend on a knowledge of the context, but each sentence in itself should be intelligible to the reader.

The beginnings of letters produce many mistakes of this kind:

In reply to your enquiry about maintenance.
With reference to your telephone call.

These are incomplete parts of sentences (in reply to your enquiry... what?) and do not make sense by themselves. What is more, they do not contain verbs. They are not sentences.

A sentence should contain one or two items of information and not twenty-three. Engineers often feel unable to stop writing, and add sentence to sentence, separating sentences by commas, until they have moved far away from the first piece of information, as in the following example:

When you have logged into the computer and typed in
your user name, the computer will respond with a dollar
prompt, you can then set your password by invoking the
'passwd' command, this password should be used
whenever you log in in future.

This is not easy to read and, what is worse, not easy to carry out. There is one sentence which ends with 'prompt' (and a comma!), a second sentence which ends with 'command' (and another comma!) and a third sentence which ends with 'future' (and a correct full stop). As soon as these units of information are separated, the passage is easy to follow:

> *When you first log into the computer, type in your user name and the computer will respond with a dollar prompt. You can then set your password by invoking the 'passwd' command. This password should be used whenever you log in.*

The various stages are now clearly defined, and incidentally the unpleasant 'in in' at the end has been removed. 'First' in the first sentence makes 'in future' unnecessary.

Sentences are made up of two kinds of unit: clauses, which contain a verb, and/or phrases, which don't. In each sentence, the essential element is called the main clause: this may be just one word long, as in 'Stop!', which is acceptable as the single word is itself a verb. Usually there are several words in the main clause, as in:

> *The car refused to start.*

This main clause makes good sense, contains one main idea, and has a complete verb, 'refused'. (We may question the word 'refused', on the grounds that the car had little choice in the matter, but the expression is widely used and so, perhaps, acceptable.) In other words, the main clause and the sentence are one and the same thing. However, such a sentence can be extended, for instance by a phrase, a group of words which do not contain a verb.

> *On a cold, damp morning, the car refused to start.*

We still have the main clause with its verb, but there is now additional information in the form of the phrase 'on a cold, damp morning'. It is worth noting that the phrase could occur in the middle of the main clause without affecting the structure of the sentence, as in:

The car, on a cold, damp morning, refused to start.

In this type of construction, the intervening phrase is usually separated from the rest of the sentence by a pair of commas, as in the example. A phrase may be placed at any point of the sentence, even at the end:

The car refused to start, on a cold, damp morning.

The reader will probably have noticed that there is a slight shift of emphasis: when the phrase comes in the middle of the sentence, it has more stress than at the end; in the earlier example, the weather is particularly worthy of comment, while the problem of the car is stressed more heavily if the description of the weather is placed later.

Phrases are often, though not always, descriptive of time or place: examples are *in the afternoon, after working hours, at the same time, at the company's headquarters,* and so on. They add to the meaning of the rest of the sentence, but as they do not contain a verb, they can never stand alone.

A sentence, then, will contain a main clause and may contain one or more phrases. It may also have other clauses, subsidiary clauses, which include verbs but which cannot, unlike the main clause, stand alone. These clauses expand the meaning of the main clause, often explaining how or why the main action was taken. For example, the writer may want to explain why the car refused to start, and therefore adds the subsidiary clause 'because the battery was flat'. There may be some resentment about the inconvenient timing of the incident, and the writer adds another subsidiary clause, 'when I was already late for work'. The sentence now reads as follows:

On a cold, damp morning when I was already late for work, the car refused to start because the battery was flat.

This sentence is still acceptable, because all the information is linked to the original idea, but the reader may well feel that the limit for easy reading is near. Trouble arises if more ideas are added:

> *On a cold, damp morning when I was already late for work, the car refused to start because the battery was flat, but my neighbour, who saw the problem, came out with jump leads, and both of us together started the car and I got to work only ten minutes late.*

If we analyse this sentence, we find that it contains four main ideas:

1. *the car refused to start*
2. *my neighbour came out with the jump leads*
3. *we started the car*
4. *I got to work*

Each of these ideas has its subsidiary information:

1. *on a cold, damp morning* (phrase)
2. *when I was already late for work* (subsidiary clause)
3. *my neighbour* (phrase)
4. *who saw the problem* (subsidiary clause)
5. *both of us together* (phrase)
6. *only ten minutes late* (phrase)

None of these items by itself is a sentence (none makes sense by itself), but if each is added to the appropriate main idea (main clause), a sequence of good, readable sentences results:

1. *On a cold, damp morning when I was already late for work, the car refused to start.*
2. *My neighbour, who saw the problem, came out with jump leads.*
3. *Together, we started the car.*
4. *I got to work only ten minutes late.*

Each sentence now fulfils the criteria given in the definition, and we have a structured piece of writing, with all the events in the correct logical order.

One more aspect of sentence structure is worth mentioning at this point. As we have seen, a sentence may have one major idea, in which case it is called a 'simple sentence'. It is also possible for a sentence to have two or more major ideas of equal importance (a 'compound sentence'), although the writer must take care not to get carried away by this possibility. The ideas must be closely linked and none must involve many words, or the sentence will become too long. The essential element of such a sentence is that there are words which make the connections – the ideas must not simply be placed one after the other with a comma in between, which is a common mistake in technical writing. The two words which most frequently act in this way are *and* and *but*; in each case they join together main clauses which could stand independently. Such joining words are known as 'conjunctions'.

As an example, we might well decide, for reasons of style, to join together sentences 3 and 4 of the example given above:

Together, we started the car and I got to work only ten minutes late.

If we do so, then we have made a *decision* on the grounds of good style; we have not simply allowed the sentence to happen.

Understanding the basic structure of a sentence is an important stage in developing a good style. It gives the writer considerable power, both to show a clear logical sequence of events and to stress specific ideas. The reader perceives that the writer is in command of the information and has presented it in a structured way. The reading is as easy as the content allows – provided that the sentences so constructed have not been allowed to grow out of control.

Sentence Length

A sentence, especially if it contains technical information, should not be so long that the reader is unable to assimilate the ideas. An average of seventeen to twenty words is reasonable, and a maximum of about forty words is sensible. Even if a sentence is correctly constructed, it will be difficult to untangle if it contains too many ideas (the 'Sellotape effect'), as we can see from the following example:

> *Further to our recent meeting regarding electricity supply and utilisation, I would like very much to arrange a further meeting with you to discuss the subject, coupled with a general discussion on electrical applications and equipment capable of providing possible reductions in unit production costs, such as electric/steam generators and convection/radiant ovens coupled to load control equipment.*

This amazing sentence contains fifty-nine words and several ideas. Underneath it all is one simple unit (the main clause which is itself a complete sentence):

> *I should like to arrange a further meeting.*

Two subsidiary ideas are added to the beginning of this:

> *1.   further to our recent meeting*
> *2.   regarding electricity supply and utilisation*

and one subsidiary idea is added to the end:

> *3.   to discuss the subject*

This is quite enough for one sentence. However, our engineer sweeps on to the next basic idea:

> *We could have a general discussion.*

To this, yet another subsidiary idea has been added:

1.   *about electrical applications and equipment*

and, to make matters worse, further details follow:

2.   *capable of providing possible reductions in unit pro-*
     duction costs

Refusing to give up, the writer moves into examples:

3.   *such as electric/steam generators and convection/radi-*
     ant ovens coupled to load control equipment

The sentence is now a very long way from the meeting which was to be arranged, and the reader is thoroughly confused. There are three main ideas in this monstrous sentence:

1.   *I should like to arrange a further meeting.*
2.   *At the same time, we could discuss equipment.*
3.   *I can give you examples of the kind of equipment I*
     have in mind.

Subsidiary ideas can now be grouped round each of these main ideas to form three sentences. In organising the information in this way, the writer might well decide that the second and third sentences belong in a new paragraph (see Chapter 4), and to leave out redundant material such as 'to discuss the subject'; the pompous word 'utilisation' may be replaced by the simpler 'use'. The details could then be reorganised to produce a more logical and readable version:

I should like to arrange a further meeting with you to
continue our discussions of [date] on electricity supply and
use.

At the same time, we could have a general look at
electrical applications and equipment [which are] capable
of providing reductions in unit production costs. I have in
mind electric/steam generators and convection/radiant
ovens coupled to load control equipment.

Sometimes, as we have seen, a sentence contains only one basic idea, while other sentences contain more than one basic idea, but the guidelines for sentence length always apply.

☐ Sentences contain one idea, or two or three closely related ideas which must be correctly joined together.

In passing, it is worth noting that *however* is not a conjunction, although it is often wrongly used as one. It either comments on the information:

> *The initial cost of the machine is high. Maintenance, however, is relatively inexpensive.*

or it is the equivalent of 'in whatever way':

> *However we look at the problem, there is no easy solution.*

We read these two sentences in different ways because of the punctuation (the two commas) in the former, and the absence of punctuation in the latter (see also p. 52).

Good style includes variety of sentence length. A few short sentences are direct and sometimes dramatic in their impact, as, for instance, the seven-word sentence which begins this paragraph. Too many short sentences, however, give a rather childish effect, as if the writer thinks that the reader will have difficulty with more complex sentences. It is better to look for logical connections between some of the sentences, and to join them in an appropriate way.

Perhaps the most important rule for good technical writing is to avoid over-long and over-complicated sentences. Many other problems, especially those connected with punctuation and grammar, will disappear if sentence length is controlled. A sentence of more than about forty words causes two areas of difficulty:

1. *for the writer,* who finds it hard to organise the construction of the sentence if there are too many ideas to communicate at once;

2.   *for the reader,* who finds it hard to assimilate the
information in a long sentence, however well it is
written.

The following sentence is not easy to read; it makes no
concession to the reader:

*For higher power applications chopper drive which is
more efficient but more complex than dual voltage drive
which requires two supplies must be used although it is
complex, generates audible noise at the chopping
frequency possibly causing interference and additional
iron losses in the motor.*   (46 words)

We can analyse the problems as follows:

1.   The sentence is too long, and contains too much
information. It must be divided up.
2.   There is a very long digression in which chopper drive is
compared with dual voltage drive. Fifteen words separate 'chopper drive' from 'must be used'.
3.   Within this digression, another independent sentence is
begun, which contradicts the message of the whole passage. Careless reading might easily pick up 'dual voltage
drive which requires two supplies must be used'.
4.   There is a floating 'it'. When the reader meets 'it is
complex' and looks back to see what 'it' refers to, it is
easy to assume that 'dual voltage drive', the nearest
singular noun, is meant. In fact, 'it' refers to 'chopper
drive', now twenty words away. The problem of 'it'
continues through the next verb, 'generates', com-
pounding the confusion.
5.   'Causing' has no clear subject. It appears to belong to
'audible noise' but this does not make sense. Similarly, it
could refer to 'two supplies' or, of course, 'dual voltage
drive'. Again, the intended message is contradicted: the
subject of 'causing' is in fact 'chopper drive'.

The passage can now be rewritten in a more acceptable
form:

For higher power applications, dual voltage drive or chopper drive must be used. Dual voltage drive is simpler than chopper drive but is less efficient and requires two supplies. Chopper drives are very efficient but complex, generating audible noise at the chopping frequency and possibly causing interference and additional iron losses in the motor.

The two options are introduced at once, and then each is considered in turn. There are now three sentences, of thirteen, sixteen and twenty-five words, giving variety without the confusion of an over-long sentence. The writer is clearly in control and the reader can assimilate the information with ease.

Variety of sentence length within the document can be used by the writer to control the reader's approach. Straightforward facts can be written in comparatively short sentences, perhaps up to twenty words, which can be read quickly. Passages which require more consideration, such as the conclusions of a report, often call for longer sentences, perhaps up to thirty-five or even forty words, to slow the reader down and command concentration. Readers are unlikely to notice such manipulation.

- Sentences must not be too long: forty words is a sensible maximum.
- Variety in sentence length helps the reader.

Sentence Construction

We have already looked at two kinds of sentences: those which express one basic idea (simple sentences), and those which express two or more closely related ideas (compound sentences). Some compound sentences can be made up of equal basic units (main clauses) joined by *and* or *but* (*never* by a comma alone), while others consist of a main unit and further units which are of lesser importance (subsidiary clauses, or phrases).

Knowing the various possible ways of constructing a sentence allows the writer to make decisions. What aspect of the information is to form the main clause of the sentence? The emphasis of the message will depend on the writer's choice. In the sentence

The machine was overhauled, after which it worked at full capacity.

the main clause is clearly 'The machine was overhauled', and this will be recognised as the most important aspect of the information. If the sentence is changed round, so that we read:

The machine worked at full capacity after it was overhauled.

the emphasis is now firmly on the fact that the machine worked at full capacity, since this is the main clause of the sentence.

This ability to move the emphasis within the sentence is a useful one. We can see the difference between the 'neutral' sentence

The machine was badly damaged but it could be repaired.

and two versions which carry different weight:

Although the machine was badly damaged, it could be repaired.

in which the stress is on the repair, and

The machine was badly damaged, although it could be repaired.

in which the stress is on the damage.

The reader must always be able to identify the main unit of the sentence and to recognise where the emphasis has been placed. It is possible for a writer to be so involved with subordinate units that the main unit of the sentence is accidentally left out, as in the following:

In order to prevent damage to integrated circuits as a result of static discharge which may be caused by a variety of factors such as incorrect handling procedures, poorly grounded instruments or even on occasion by electrostatic buildup on clothes, shoes or floor coverings.

If we try to pick out the main unit of this sentence, we shall soon find that it hasn't got one. 'In order to prevent damage...' what must be done? We are not told. Again, we have a sentence more than forty words long, which is completely out of control. As soon as we identify the main idea in the information and give it a separate sentence:

Damage to integrated circuits may be the result of static discharge.

the rest falls into place:

This may be caused by incorrect handling procedures, poorly grounded instruments or electrostatic buildup on clothes, shoes or floor coverings.

Now that the two sentences have been planned, the redundant details are recognised as such, and removed. 'A variety of factors such as' and 'even on occasion by' add nothing to the meaning; when they have been taken out, we are left with two grammatically correct sentences of eleven and nineteen words, an acceptable length.

□ The main unit (clause) of a sentence conveys the main idea, and must be readily identified by the reader

When the main unit is so identified and the subordinate units are planned, there are still two possible ways in which the sentence as a whole can be organised. The subordinate ideas can be put first, so that they lead up to the main idea. This is a useful technique in writing thrillers, as the tension mounts until the climax is reached:

> *As night fell, and with it the hard, driving rain, his determination increased and, breathless, exhausted, running onwards although he no longer knew in what direction, stumbling and nearly falling so that he grazed his hands and twisted his ankle, he now knew more certainly than ever that he would, if need be, die rather than surrender.*

Everything in the sentence leads up to that final, dramatic statement that 'he would die rather than surrender', and the impact is made stronger by the buildup of detail: the night, the rain, his exhaustion, loss of sense of direction, pain – *despite all this*, the writer is saying, death was preferable to surrender.

Such is the art of the novelist; the engineer writing technical information must approach the material very differently. The reader must know *at once* what the writer is talking about, so that all the subsidiary ideas can be weighed and considered. Tension is irrelevant. A 'back-to-front' sentence, with the main point at the end, is merely irritating:

> *To present detailed waste water proposals in the absence of community structure plans is not possible.*

Only at the end of the sentence does the reader get the message: it is not possible. If the sentence is turned round, it becomes clear at once:

> *It is impossible to present detailed waste water proposals in the absence of community structure plans.*

There is an additional advantage in this ordering of the sentence. 'Present' can in writing, although not in speech, be either a verb ('to presént') or an adjective ('présent'), and we do not know how to interpret it until we understand the context. In the original form of the sentence, we cannot be sure that 'presént' is intended until at least half-way through; in the inverted form of the sentence, we know almost at once.

There is another reason for organising sentences the 'right way round'. The writer wants to attract the reader's attention, and it is unhelpful to leave the most appealing information until the end:

> *Further to our discussions and my brief survey, I have pleasure in giving details below of my findings together with approximate indications of the likely savings that might accrue by raising hot water directly in each office at approximately 100% efficiency and saving distribution costs.*

The engineer who wrote this real-life sentence had an important message: I can increase efficiency and save money. Who could resist the appeal of such a sentence? This exciting offer is totally hidden in words, buried at the end of a long sentence of forty-four words instead of being highlighted:

> *In the light of our discussions and my brief survey, I have pleasure in giving my findings.*
> *Savings can be made and efficiency increased by raising hot water directly in each office.*
> *Details are as follows:*

Three ideas are presented: the findings result from discussions and a survey; savings can be made; detailed information follows. The interesting word 'savings' now appears at the start of a sentence, where it is likely to catch the reader's attention.

□ Start the sentence with the most important idea.

Of course, this is a guideline rather than an absolute rule. Occasionally, for the 'political' reasons which may affect an engineer's writing, it might be better to write a 'back-to-front' sentence:

> *Your proposal is interesting... in many ways attract-*
> *ive... and if the economic climate were different... but*
> *we must regretfully decline...*

The approach is gentle to make the refusal more palatable!

So far, we have looked at sentences which have a main clause or main clauses and subsidiary units, either phrases which have no verbs, or subordinate clauses introduced by words like *when, although* or *because.* It has been assumed that the verb can be easily identified, and that it consists of a single word. This is not always so. Verbs can be made up of more than one word: I *am working,* they *were working.* Both the words in italics are needed if the main clause of the sentence is to make sense. Both are parts of the main verb.

The verb *am working* is made up of two parts, an auxiliary verb (*am*) and a participle (*working*). The auxiliary verb does the work of letting us know the person and the tense of the verb, for example whether I *am* or they *are,* and whether I *was* or they *have been.* The participle, which often ends in 'ing', tells us what action is being taken ('working' rather than 'sitting' or 'writing'). As soon as one part of the verb is missing, the sense is lost. So, for example, we cannot say *I working on the night shift* any more than we can convey the meaning by saying *I am on the night shift* (the reader doesn't know what I'm doing – I could be sleeping!). If the verb consists of more than one word, every word has its part to play in giving the message.

However, if a sentence already contains a complete main verb, in practice we often use just a participle, understanding the missing auxiliary verb. 'Working with the lathe' is clearly neither a main clause nor a full sentence, as it does not make sense. We could make it complete by adding to it: 'I was working with the lathe.' But the main point of the message might be different: 'I produced a smooth finish.' This is also a main clause/complete sentence. Both sentences have the same subject, I, and for this reason we might choose to bring the two ideas together by omitting 'I was' from the first sentence. Thus we get:

Working with the lathe, I produced a smooth finish.

This is a correct sentence, with the main unit 'I produced a smooth finish', and a subordinate unit, 'working with the lathe' which does not by itself make sense. We can understand it because the word 'I' is the subject of both the participle and the main verb: *I* am working and *I* produced a smooth finish.

However, if the subjects of the participle and of the main verb are different, the result is nonsense:

Working with the lathe, the table had a smooth finish.

The table may have had a smooth finish, but it was not, presumably, working with the lathe!

This phenomenon, known as the unrelated participle, can produce not only nonsense but also ambiguous humour:

Rusting badly though it was, Jim's brain told him that he would buy the car.

Grammatically, we must assume that Jim's brain, the subject of the main clause, was rusting badly, but nevertheless told him .... We might guess that it was the car which was rusting (but who can be sure?). It is all too easy to produce such sentences:

Used for long periods without ventilation, overheating can cause damage to the instrument.

Is overheating used for long periods, as the sentence says? The intended meaning was probably

Overheating can cause damage to an instrument which is used for long periods without adequate ventilation.

The auxiliary verb 'is' has now joined its participle 'used', and the sentence makes sense.

☐ Every sentence must have a complete main verb.

A similar problem can arise with the infinitive of the verb (the 'name' of the verb, such as *to be, to work, to eat*). The sentence

> *To drive well, your eyesight must be good.*

suggests that your eyesight can drive, well or otherwise. This is clearly nonsense, and so the sentence must be changed:

> *Your eyesight must be good if you are to drive well.*

The verb 'are to drive' is three words long, but all are needed if the sentence is to make sense.

Infinitives are best known for being split. Good style frowns at a word straying between 'to' and the verb: *to effectively control* is bad, while *to control effectively* is good. Generally this is true, although, as with many English rules, this one can be broken *if the writer knows why it is being broken*. Emphasis is thrown on the apparently misplaced word, and such emphasis may be justified:

> *To safely remove the radioactive material . . .*

This split infinitive stresses the important word 'safely', and so may be acceptable. Such a device will lose its effectiveness if it is used too often; very occasionally and for a clear purpose, it makes its point.

Some of the problems which we have been looking at result from the writer's unwillingness to look at a sentence as a whole.

> *The run time checks indicate at what point of the process the computer is currently at.*

This writer has changed constructions in the middle of the sentence, so that it falls apart. It could be written either as:

> *The run time checks indicate at what point of the process the computer currently is.*

which is awkward, or as:

The run time checks indicate what point of the process the computer is currently at.

This is acceptable, although purists might quibble because the sentence ends with 'at', which sounds clumsy. The difficult word is perhaps 'is', which conveys very little in this sentence. If the writer thinks about the intended message, it could end up as:

The run time checks indicate what point of the process the computer has reached.

which is easier to read and removes the need for 'currently'.

It is always useful for the writer to think of the message in terms of: If I had the reader in front of me now, what would I say? The written language is more formal than the spoken language (see Chapter 5 on good style), but once the message has been identified, it can be 'tidied up' for the written form.

I hope to be able to confirm the appointment of James Twigg within the next few days and can assure you that subject to your response to my questions being positive, immediate liaison will be set up between you so that he may be instrumental in arriving at the appropriate conclusions which will benefit us both.

What was the author of this amazing sentence trying to say?

As soon as we've confirmed James Twigg's appointment, which shouldn't take more than a few days, and as long as you're happy with our suggestions, we'll put you in touch with each other. With luck, you'll get on and we'll all do well out of it.

That is too chatty for a written message, but it can be made a little more formal without becoming pompous:

We hope to confirm James Twigg's appointment within a few days. As soon as you have agreed to our suggestions, he will contact you to establish a good working relationship.

The two essential units of information are:

1. J.T.'s appointment
2. J.T.'s making contact

It is obvious that a good working relationship is of benefit to both parties, and there is no point in saying so. A fifty-five word sentence is replaced by one sentence of eleven words and another of nineteen words – only thirty words in all.

Sentences which ramble, like the one above, can accident-ally give misleading information. Again, the essential mes-sage must be identified and made clear to the reader.

The lower temperature of 30 degrees Celsius is taken to be the typical external ambient temperature for the equipment in normal use whilst the upper one, 80 degrees Celsius, represents the maximum operational external ambient temperature.

The catch in this sentence is that it mentions 'lower' and 'upper' temperatures, and 'upper' is followed by the word 'maximum'. A careless reader might easily assume that 30 degrees is therefore the minimum, rather than the typical, temperature. The message has to be re-formed without ambiguity (and with 'represents' replaced by 'is'):

Typically, the equipment will be used at an ambient 30°C. The maximum operational ambient temperature is 80°C.

The key words which are contrasted, 'typically' and 'max-imum', are now highlighted at the beginning of the sentences so that no confusion is likely. The temperatures themselves have been abbreviated in a generally recognised way as '30°C' and '80°C'. Each appears at the end of its sentence, so that again the contrast is clear. The unnecessary information 'lower' and 'upper' has been removed, as has the word 'temperature' in the first sentence (what else would 30°C be?). In the second sentence it is retained for ease of reading.

Redundant words and phrases are common in technical writing. Expressions such as *Initially, we began by...*, *a new innovation*, *future consequences* (or, more subtly, *the consequences which lie ahead...*), *clearly obvious* or *both the two rivets* all contain unnecessary words. Perhaps the writer is lacking in confidence and feels the need to say the same thing twice. If so, such a message should not be conveyed to the reader. Not only does it suggest hesitation, but it makes the document longer than it need be, which is a waste of time and money.

☐ Write as concisely as possible; avoid unnecessary words.

There are other difficulties with sentence structure which also arise from a lack of planning. A sentence may move from the singular to the plural, or *vice versa*, because the subject is not clearly identified:

> *The failure of the systems which we have installed recently have led to the current financial crisis.*

The subject of the main verb in this sentence, 'have led', is the singular word 'failure': the form of the verb should therefore also be singular, 'has led'. We can see how this mistake arose. Between the singular word 'failure' and the plural verb 'have led', there is the plural word 'systems' followed correctly by the plural 'we have installed'. Two instances of the plural have led the writer to assume that the verb needed must also be plural, forgetting that the original subject word 'failure' was in fact singular. As a check, the writer should ask, '*what* led to the current financial crisis?' to which the answer is, 'the failure'.

The word 'each', which is singular, can produce similar difficulties.

> *Each of the systems which we have installed recently have given us initial problems.*

The subject of 'have given' is each, meaning 'each one', and the sentence should thus read:

Each of the systems which we have installed recently has given us initial problems.

□ The subject and the verb must agree, both singular or both plural.

Two other dangerous subject words are *it* and *which*. Both are easily attached to the wrong word or phrase; sometimes the reader finds great difficulty in relating them to anything:

The question of how this work should be carried out is one which it is difficult to answer. During our discussions, I said it might be possible to work from a cradle but due to the instability of this type of apparatus, it could prove extremely difficult. Also, due to the large scale of work involved, it would increase the length of the contract.

There are four uses of 'it' in this passage, and none is easily linked to a meaning. Much of the first sentence is redundant in any case; it means no more than 'we are not sure'. In the next sentence, the second 'it' means little more than 'this', and the sentence could be reorganised into a more concise form. The third sentence suffers from the same problems as the second. The passage could be rewritten as:

We are not yet sure how to carry out this work. Using a cradle, as I suggested previously, would be difficult because of its instability. Given the scale of the work, it would be time consuming to use this method.

'Which' is also tricky:

Manuals are mainly held in the print room, but some by individuals which are often unique.

Grammatically, 'which' refers to 'individuals', but there are two problems here:

1. individuals are *always* unique;
2. individuals are people, and therefore referred to as 'who' rather than as 'which'.

Commonsense, but not grammar, tells us that 'which' refers to 'manuals', and that the sentence should be rearranged to read:

> *While most manuals are held in the print room, some which are unique [presumably, 'which are the only copies'] are held by individuals.*

An expression much favoured by engineers is 'such that'.

> *Data have been collected from field surveys such that future projects can be planned.*

(For a discussion of the word 'data', see p. 17.) In this sentence, 'such' is ambiguous. It may mean 'in such a way that':

> *Data have been collected from field surveys in such a way that future projects can be planned.*

In this case, it is the *method* of collection which has enabled the planning of future projects. However, 'such' may mean 'so that':

> *Data have been collected from field surveys so that [= in order that] future projects can be planned.*

Here, data were collected and *as a result* it is possible to plan future projects. 'Such that' can, of course, be used correctly:

> *The method of collection was such that it enabled future projects to be planned.*

In this case, 'such that' means 'of such a type' or 'of such a quality'. This use is comparatively rare, and it is worth checking that 'such that' has not been used when a different meaning was intended.

Perhaps it is a natural diffidence among engineers which encourages them to write in a negative way. Expressions such as *it is difficult to deny, it is not unlikely that,* and *it is not possible to disprove that* suggest that the writer is hesitant (or diffident). Carried to excess, such writing ends up by undermining the reader's confidence:

> *If the trend shown continues, then there should be no reason why an improvement in productivity of approaching 40% is not achievable.*

This is good news, but the reader might be forgiven for feeling that the outlook is gloomy. 'No reason' followed by 'not achievable' leaves a confused impression. The writer was trying to say

> *If we keep up the good work, we'll just about make a 40% rise in production.*

or, more formally,

> *If the present trend continues, we should approach a 40% rise in production.*

or, to give emphasis to the good news,

> *A 40% rise in production is forecast if the present trend continues.*

which sounds much more cheerful than the original sentence!

Perhaps negatives are most frequent when the news is bad:

> *It is impossible to deny that if the present trend is not reversed, production figures will fail to show the expected rise of 40%.*

The writer crowds negatives together (impossible, deny, not reversed, fail to show) presumably so that the reader is led gently to the dismal tidings. Such writing can produce very difficult statements:

It is impossible to deny that production figures have not risen.

Presumably this means that we have to admit the truth, but we would much rather not do so.

☐ Plan the sentence so that it says what it means in a positive way.

As a final challenge to the readers of this chapter, we have two 'real-life' passages in which the meaning is somewhat obscure, not because of the technical content but because of faults in the writing. Simplified versions follow.

Original versions of these passages
    1. *Particular care should be taken to ensure that where an activity has overrun, then that resource requirement for the remainder of that activity is reflected in the current report in addition to the resource needed to maintain the programme during the reporting period.*
    2. *It should be noted that the project group's decision to resume work on the existing prototype in no way indicates any belief in non-feasibility of the other possible methods. The reason for carrying on with the original design being mainly a practical one in that it would seem to be more worthwhile attempting to successfully terminate one approach to the topic rather than start again from scratch using another method and possibly only, with luck, reaching prototype stage.*

Simplified versions of these passages
    1. *(Please) note that reports should show both the resource requirements for any activity which has*

> *overrun, and the normal requirements for the cur-*
> *rent period.*
>
> 2.   *The project group has decided to resume work with*
>   *the original design, since otherwise it might not be*
>   *possible to get beyond the prototype stage.*

Punctuation

This section deals with the most commonly used punctuation marks, and gives simple guidelines for their use.

Full Stops, Exclamation Marks, Question Marks

Full stops are found at the ends of sentences and, although
less frequently nowadays, in abbreviations. The sentence
which you have just read shows the conventions: a capital
letter at the start and a full stop at the end. Engineering
writers tend to overlook full stops and to use commas instead
(see p. 24). This is always wrong, as the following example
shows:

> *Sentences cannot simply be put together, they need a*
> *word or phrase which joins them, they should otherwise*
> *be separate, the commas should be changed to full*
> *stops.*

This is a simple example of a common problem. The writer is
saying:

> *Sentences cannot simply be put together.*
> *They need a word or phrase which joins them.*
> *They should otherwise be separate*
> *The commas should be changed to full stops.*

The four sentences above are correct, but dull. Bearing in
mind what was said earlier about variety of sentence length

(see p. 31), it seems appropriate to use a joining word (conjunction) in one case:

> *Sentences cannot simply be put together. Either they need a word or phrase which joins them or they should be kept separate. The commas should be changed to full stops.*

Either... or is a good way of joining two sentences which obviously show alternative ways of tackling the sentence problem.

The alternatives to full stops are exclamation marks, question marks and semicolons (which are dealt with separately). Exclamation marks, as their name suggests, show sudden reactions: *Stop! Look out! That's amazing!* Their use in engineering writing is, to put it mildly, rare.

Question marks, again as one would expect, show the exact words spoken in a question:

> *How should we tackle this problem?*
> *What results might we expect?*

Such questions are usually unhelpful in writing (although often useful in speech) as the writer is about to give the answer. Writers are sometimes tempted to use questions as headings:

> *Why study engineering nowadays?*

Again, this is not helpful to the reader, and is usually considered to be poor style. The only person who is going to answer the question is the writer who asked it, and so why bother? Occasionally, a series of questions can be used to draw attention to what the reader would want to ask, each question being followed by an appropriate answer:

> *How does the machine work?*
>   *It works by...*
> *How much does it cost?*
>   *The basic cost is...*
> *What about maintenance?*

A thorough overhaul has to be carried out...

As the above example suggests, this technique is usually applied in sales literature; it is not appropriate in, for example, a technical report.

Full stops used to appear in abbreviations, either of the 'name' variety (for instance, U.N.I.C.E.F.) or of the Latin original variety (for instance, i.e.). This is now rare although it is still correct; the modern convention allows the initials to run together (UNICEF; ie). If there is any chance of misunderstanding, then it is better to write the term in full (as in the case of litres, when an abbreviation to the single 'l' can be misread). If a company or organisation maintains full stops in its title, then of course they should be used.

Semicolons

Semicolons may also take the place of full stops. If two sentences are closely related in subject matter, perhaps by contrast or by a common factor, the relationship can be stressed by joining the sentences with a semicolon:

The design of the bridge was superb at the time; today it has to carry too much heavy traffic.

Each 'sentence' remains an accurate, grammatical whole, making sense by itself. However, the link of information (then...now) is strong, and is emphasised by the use of a semicolon instead of a full stop at the end of the first sentence (followed by a small, not a capital, letter). This use of the semicolon can produce an elegant style; if it is not overused, it can be most effective.

Semicolons can also be used to separate sections of information in a list:

The following hazards must be considered:
(1)   insulation and protection from electric shock;

(2)   fire risks and the location of fire extinguishers;
(3)   testing of pressurised or other highly stressed
        components.

Nowadays, these semicolons are often omitted, which is acceptable as long as the individual items in the list are short; if they are more than a line in length, then for clarity the punctuation should be retained.

Colons

Colons are not interchangeable with semicolons. The most common use of a colon is to introduce an example, as we have seen in this book, or to introduce a quotation, as in the case of the following extract from a wise article about the importance of identifying the readers' needs when writing technical documents:

> *Most documentation is incomplete because it tells the reader precisely what to do but provides no contextual information on why each step is required, useful, or efficient. The result is that users are more likely to learn how to perform a function than how to link it to other functions to perform more complex tasks; without the documentation, many users will be unable to perform the function because they can't remember the next step.*[2]

A colon may also introduce a list, as in the following example:

> *The equipment needed for this test is as follows:*
>
> *oscilloscope*
> *digital voltmeter*
> *signal generator*
> *logic analyser*
> *power supplies*
> *soldering iron*

Note that as the items in this list are all very short, they are not followed by punctuation.

A list may sometimes simply be an amplification of the preceding information:

> *There are three main types of stepping motor: variable reluctance motors, permanent magnet motors and hybrid motors.*

This is a short list, not printed in a list format and containing only three items, but the first eight words serve as an introduction to the items which follow.

Colons used to be followed by a dash (:–), but the dash is usually omitted nowadays.

Commas

In some ways, commas are a difficult form of punctuation, because while they often follow rules, they are also to a certain extent the result of individual choice, of a 'feeling' for the language.

The most common use of a comma is to separate the main part of a sentence from a subordinate part (see p. 26), either to make the meaning clear or to allow the reader to 'take breath' naturally. This is an important aspect of the use of commas, as the 'natural pause' in a long sentence helps the reader to assimilate the information given so far and to prepare for what is to come. The division within a sentence can be clearly seen in the following:

> *When the bridge was first built, it was adequate for traffic requirements.*

If we read this sentence aloud, we shall pause naturally at the end of the subordinate unit (after the word 'built') before moving on to the main part of the sentence. The sequence can, of course, be reversed:

The bridge was adequate for traffic requirements, when it was first built.

We may not feel the need for a comma after 'requirements' (the 'feeling' for language), but as soon as the subordinate unit becomes longer, we need the comma again to allow us a brief pause:

The bridge was originally adequate for traffic requirements, but today there are frequent holdups and sometimes long queues stretching back towards the motorway.

The subordinate unit, as we have seen, may appear neither at the beginning nor at the end of the sentence, but in the middle:

The bridge, originally adequate for traffic requirements, is today the scene of frequent holdups and the cause of long queues.

The phrase 'originally adequate for traffic requirements' is now enclosed by two commas, so that it is separated from the main sentence, which would, indeed, make sense without it. Both commas are necessary; if one is left out, the sentence will not read correctly.

We have just seen a different but related use of the comma:

The main sentence would, indeed, make sense without it.

In this case, one word, 'indeed', is between commas. It is a comment on the rest of the sentence and could be left out without a change of meaning. Many expressions such as *on the other hand, nevertheless, in spite of* . . . can be used in this way:

There is, however, a plan to build a second bridge over the river.
Nevertheless, for the time being the problem will remain.
It is possible to avoid the long queues, however.
The detour needed, it must be remembered, is lengthy.

From these examples, we can see that the 'comment' words and phrases are separated from the rest of the sentence by a comma or commas, depending on where the comment is placed in the sentence.

There is a modern tendency to leave out one or both commas, with ambiguous results:

> *The bridges, which cross the river, are in urgent need of repair.*
> *The bridges which cross the river are in urgent need of repair.*

The first sentence means that *all* the bridges cross the river, while the second sentence means that *only* the bridges which cross the river are in need of repair (those over the railway line are in good order). The test for the writer is to read the two sentences out loud, and the natural pauses of the speaking voice will make the difference of meaning clear. In writing, we have to depend on the presence or absence of the commas to show us which is meant.

One more case of 'comment' words or phrases is worth noting. Sometimes in introducing a person, or a book, article, and so on, we add an explanatory comment:

> *James Twigg, an engineer who works locally, came to the meeting.*
> *James Twigg, Chairman of the local branch of the Institution, came to the meeting.*
> *James Twigg's book, which has been so well reviewed, will be on sale after the meeting.*
> *'Modern Bridge Design', by James Twigg, is on sale now.*

The same rules apply as for the other comment words and phrases: a comma before and after separates the comment from the rest of the sentence.

Commas can also be used to separate short items in a list, when they are written within the text rather than listed down the page:

In digital electronic hardware testing, the engineer makes use of an oscilloscope, a dvm, a signal or pulse generator and a logic analyser.

There is not usually a comma before the 'and' at the end of the list (but see below).

The main uses of commas in technical writing have now been discussed. The remaining uses are less important, but it is worth noting that commas are used before or after direct speech:

He said, 'I shall cross the bridge.'
'I shall cross the bridge,' he said.

There are also the commas which are placed to help the reader, for instance, the comma before 'and' (usually forbidden in school!). If the sense is made clearer by the comma, then it should be used:

I crossed the bridge and the mountains lay all before me.

The reader might well understand, 'I crossed the bridge and the mountains', and then have to readjust the reading as it becomes clear that the sentence continues. A comma makes all plain:

I crossed the bridge, and the mountains lay all before me.

Sometimes, two or more uses of 'and' in a sentence have different values:

I crossed the flat marsh land and then the bridge and at last the vast and beautiful mountains came into sight.

This sentence contains three versions of 'and'. The first and the third are comparatively unimportant, linking related ideas. Indeed, 'vast and beautiful' is almost to be understood as one word. The voice's natural pause comes after 'bridge', and that is where the comma should be:

> *I crossed the flat marsh land and then the bridge, and at
> last the vast and beautiful mountains came into sight.*

The effect of leaving out such commas can be ambiguous:

> *I went to the meeting with Jim and Sarah and Peter came
> later.*

When this sentence is spoken, the natural inflection of the
voice makes the meaning clear, but when we read it we have
no idea whether I went with Jim and Sarah (Peter was late) or
whether I went with Jim only (Sarah and Peter were both
late). A comma reinforcing the important 'and' (that is, after
either 'Jim' or 'Sarah') makes the sentence unambiguous. It
is in fact reflecting the pattern of the spoken sentence, which
is often the job of a comma.

The voice has been used as a guide several times in this
section, and it is a good one. If in doubt, read the passage
aloud and notice where the voice pauses naturally. Mark the
place with a comma.

Quotation Marks

Quotation marks are used less than they used to be. For
quotations which are longer than one line of text, convention
indicates that the writer should start a new line, indenting the
words quoted and going back to the left-hand margin at the
end of the quotation. This style was followed in the quotation about the importance of contextual information on p. 51.
However, shorter quotations must still be acknowledged as
such, and for that purpose quotation marks are needed. The
normal practice is to use the single form round the words
quoted:

> *A useful comment about the need to use clear language is
> 'stark simplicity is the best form of shock tactics in the war
> of words'.*[3]

All quotations must be acknowledged at the end of the text (see p. 112).

Dashes and Brackets

Asides, comments or examples may be placed between dashes – as in this sentence – as an alternative to commas. Dashes tend to be informal in style and should be avoided in technical writing, especially when there is mathematical information included. It is too easy for a dash to be read as a minus sign.

Brackets, on the other hand, are 'heavy' punctuation. They break up the flow of the reading, and should be used only when the information which they enclose is not an integral part of the sentence. Notes like (see Figure 6.1) perhaps show the most common use of brackets. However, irony or personal comment may be shown in this way, as in a previous sentence about commas:

> *There remain the commas which are placed to help the reader, for instance, the comma before 'and' (usually forbidden in school!).*

Such a use is obviously very rare in technical material. In any case, too many brackets on the page look unwieldy and dull.

Hyphens

Hyphens are sometimes confused with dashes, but they are shorter (- rather than –) and have different uses. They may be used to bring together two words which gain a new meaning from being joined: *re-cover* is different from *recover*, *cross-section* is different from *cross section*. The hyphen may give emphasis to the idea of repetition (*make* and *re-make* stresses the 're' aspect), or, most importantly, it may be used to help the reader (*re-emerge* is easier to recognise than *reemerge*). This last usage has a major impact on the flow of the writing; if the reader has to readjust the reading on a regular basis, the action of reading is slowed down and made uncomfortable. The modern tendency is to omit hyphens wherever possible, and this is often carried too far. While the disappearance of the hyphen from a word like cooperate or subcontract is unlikely to cause problems, other words such as reallife simply look peculiar and are sometimes unrecognisable. Whenever the reader will read a word more readily because it has a hyphen, the careful writer will provide one.

Many new words start by including hyphens, and then by convention become one word, such as *online, bandwidth, wordprocessor* (this last being seen also as two separate words, *word processor*). Such words rarely cause a problem once their form has been established. Word-breaks (an example of a useful hyphen) occur when a word is divided between two lines of text, although this is of course much rarer in wordprocessed material than it used to be in typing. However, if such a word-break does occur, a hyphen can be used to bring the join to the reader's attention, especially when a word which is divided can be read as two separate words (*rearrange* being seen as *rear range*, or *legend*, entertainingly, becoming *leg end*).

Scientific and technical terms often contain hyphens which reflect two aspects of the meaning. Such words include *bio-degradable, vacuum-sealed, three-dimensional* and *single-track*. There are many others. The meaning depends on the hyphen, and it should never be omitted.

Apostrophes

Apostrophes produce more headaches than any other form of punctuation. They are either over-used, in words which happen to be plurals ending in 's', or ignored altogether, even when this creates ambiguity.

There are two uses of an apostrophe: to show where a letter or letters have been omitted, and to show possession. The former use is rare in technical writing, as it is informal in style. It might appear in a memo or as an e-mail message:

> *Please let Jim know that I can't be at the meeting tomorrow. I've got to be on the plane for New York by lunch time.*

As a note for a colleague, this is acceptable, and the abbreviations 'can't' and 'I've' are appropriate. 'Can't' is a shortened form of 'cannot', and the apostrophe shows where 'no' is omitted. Similarly, 'ha' is omitted from 'I have', leaving 'I've'. This is the spoken language and also the informal written language. In most professional writing, such words must be spelt out in full: *it is*, not *it's*. (*It's*, the abbreviation for *it is* or *it has*, takes the apostrophe because of the omitted letters.)

The second use of the apostrophe is more difficult. It identifies the owner of an object, as in *the engineer's logbook*. It will also show if there is more than one owner, as in *the engineers' logbook*, which indicates that the logbook in question is shared by more than one engineer. Generally, in English, the apostrophe is before the 's' in the singular and after the 's' in the plural, as in this example.

There are exceptions, however. Some words do not add 's' to make the plural but use what we might think of as a different form of the word, the most common group of such words being *men, women, children, people*. These words are already plural, and so they have the apostrophe before the 's': *men's overcoats, women's shoes, children's bicycles, people's opinions*. There are other eccentric forms of plural, such as changing 'y' into 'ies', as in *secretary, secretaries*. It would be difficult to write *secretaries's*, and so the final 's' is dropped, as in *secretaries' word-processors*.

There is one group of words which suggests possession but which *never* has an apostrophe. These words follow the pattern of a verb (I have, you have, and so on):

> *my, thy, his, hers, its, ours, yours, theirs.*

'Its', meaning 'belonging to it' does *not* have an apostrophe. This is the word which causes most confusion, because of its two forms, as shown in the sentence:

It's my car which has its lights on.

The first 'it's means 'it is' and so takes the apostrophe, while the second indicates possession ('the lights belonging to it') and so does not have an apostrophe. The best way to remember this distinction is to ask: does 'it's' mean 'it has' or 'it is'? If the answer is no, there is no apostrophe. It's as simple as that!

Many writers are so uncomfortable with the apostrophe that they leave it out altogether. Usually this is not important, and it is perhaps better than having a sprinkling of unnecessary apostrophes on the page. However, occasionally the position of the apostrophe changes the meaning:

Our client's money has disappeared from the bank.
Our clients' money has disappeared from the bank.

In the first case, only one client is going to be very angry; in the second, we are going to be faced with more than one client, perhaps hordes of angry clients. Only the position of the apostrophe will show which.

Fortunately, there is a rule of thumb for deciding whether an apostrophe has been correctly placed. Write the word out, including the apostrophe. Then cover the apostrophe and everything that follows it. Ask two questions: is what is left a sensible English word, and is it what you intend to say? If the answer to both questions is yes, then the apostrophe is in the right place. For instance, if we want to write

There is a problem with the secretary's wordprocessor.

and we apply the rule, we are left with the word 'secretary', a perfectly good singular word. If that is meant, the apostrophe is correct.

If, however, the problem extended to the wordprocessors of several secretaries, we might be tempted to write 'secretarie's wordprocessors'; applying the rule would produce the nonsense word 'secretarie', which is clearly wrong. The position of the apostrophe can then be corrected to

There is a problem with the secretaries' wordprocessors.

For reasons of style, it is sometimes better to turn an expression round to avoid the apostrophe. *The book's pages* is ugly, while *the pages of the book* sounds more attractive. This inversion is often a useful device, although it is a good idea to say the expression aloud first and to listen to how it sounds. Choose whichever form sounds more natural. It may be the 'of the' form, which has the added bonus of allowing the hesitant writer to dispense with the dreaded apostrophe.

Key Ideas

- Sentences contain one idea, or two or three closely related ideas which must be correctly joined together (p. 31).

- Sentences must not be too long: forty words is a sensible maximum (p. 33).

- Variety in sentence length helps the reader (p. 33).

- The main unit (clause) of a sentence conveys the main idea, and must be readily identified by the reader (p. 35).

- Start the sentence with the most important idea (p. 37).

- Every sentence must have a complete main verb (p. 39).

- Write as concisely as possible; avoid unnecessary words (p. 43).

■   The subject and the verb must agree, both singular or both plural (p. 44).

■   Plan the sentence so that it says what it means in a positive way (p. 47).

4 Paragraphs and Format

Paragraphs: Definition of a paragraph ■ paragraph length ■ numbering ■ unity of theme ■ lists ■ organisation and layout

Format: memos □ letters □ reports □ instructions □ examples and improved versions of examples

---

Paragraphs

As we have seen, sentences contain at least one major idea (main clause, which makes sense by itself), often with subordinate aspects (clauses or phrases) added to it (see p. 24). Paragraphs are developed from a collection of ideas, all of which are linked by a central theme.

Definition of a Paragraph

□ A paragraph may be formed from a single extended idea or a series of ideas, united by theme and creating an organised and logical passage of text.

This is obviously not as rigid a definition as that of a sentence in the previous chapter: there are fewer rules about paragraph writing. Nevertheless, a text without paragraphs is difficult to read. Pages of closely printed information, with no breaks or spaces, are overwhelming, and the reader will find it almost impossible to follow a logical flow of ideas, even if it exists. The end of a paragraph allows a breathing space in which readers can make sure that a set of ideas has been understood and assimilated before they move on to the next theme. Paragraphs also serve to break up the page and so to encourage readers to read on.

Paragraphs are varied in length, according to the subject matter and the format chosen. Generally speaking, three or four paragraphs to a page look satisfactory in terms of understanding an amount of information, and of space. However, memos, letters and reports tend to have shorter paragraphs than articles or books: memos, which are short in any case, often consist of just two or three paragraphs, letters may have five or six to the page, and reports about the same, although the structure of a report (see p. 84) may dictate a different number.

There are particular problems for the writer of technical notes and articles, and it is always worth checking the format of the journal concerned. If the pages are divided into two or three columns, normal paragraphs will look very long (newspaper columns are an extreme version of this problem), and paragraphs should therefore be made shorter. Seven or eight paragraphs may look bitty on the screen, but they will be acceptable on a page which is divided in this way.

There is an apparent conflict in the last half page. Paragraph length is dictated both by unity of theme and by appearance on the page. In practice, this dual criterion is not a serious problem. Writers who are aware of the importance of presentation, can usually plan a paragraph so that it achieves unity. The very long paragraphs sometimes found in engineering reports can often be subdivided into two or three separate themes with a common thread running through them. Links between paragraphs will clarify the organisation for the reader, and are themselves a feature of good style (see p. 114).

The theme of a paragraph must be clear, and can often be expressed in a short sentence at the beginning, and perhaps also, in different words, in summing up at the end. This theme is then developed, explained and clarified by means of examples or analogies, or even undermined by the ideas expressed in the other sentences. As the writer completes the examination of one theme and moves on to the next, a new paragraph marks the transition for the reader.

In some writing, it is appropriate to number paragraphs. A letter, for instance, may deal with several topics, each discussed in a paragraph. Numbering will make the material easier to use, as the engineer who is dealing with the information can not only tick each paragraph as it is dealt with, but can also use the same numbers in any further correspondence. Some reports have a system of paragraph numbering (see p. 86), but this is helpful mostly in short reports which are discussed at meetings or over the telephone. (Some organisations demand a paragraph numbering system, in which case staff have no option but to comply; a more helpful system is suggested later in this chapter, p. 87.)

It is already clear that paragraphs have to be planned. If we take the preceding paragraph as an example, we can see that the short opening sentence contains the theme: paragraphs are sometimes numbered. The following sentence gives an example of a case in which numbering is appropriate, and the sentence after that suggests two ways in which such numbering will help the engineer. Another instance of numbered paragraphs is then introduced, and the paragraph ends with a reservation about this particular use. A new theme, the planning of paragraphs, is then introduced in a new paragraph, with the linking statement that 'It is already clear...'.

There are many different ways of organising information, but the basic guideline is: identify the theme and keep related information in the same paragraph, with the proviso that the length of that paragraph should not become unwieldy. Make the theme clear to the reader, preferably at the start, and show the transition to a new theme by starting a new paragraph.

☐ Paragraphs have unity of theme.
☐ Good paragraphing produces space on the page and encourages the reader.

In passing, we should notice that some paragraphs contain only one sentence, which is all that can be said about the subject. Letters often include such sentence-paragraphs, such as:

> *I look forward to hearing from you.*

There is no more to be said, and so these seven words form a complete, correct paragraph (see p. 78).

A list may break up a paragraph, giving more space on the page and helping the reader to assimilate the ideas one at a time. This device is particularly useful when information has to be remembered for future use (perhaps in an examination!). It would not be particularly easy to learn the following:

> *Preventive maintenance should be considered when the time interval between equipment breakdown can be predicted with reasonable accuracy, or when the cost of preventive maintenance attention is less than the repair cost when both costs include that of any lost production. It may also be appropriate when equipment failure is likely to disrupt subsequent production operations or cause customer dissatisfaction. A particularly important application is in cases when injury could result from equipment breakdown.*

The first sentence in particular is very long and contains too much information. Although the paragraph has unity of theme, the need for preventive maintenance, it is congested with detail and difficult to take in. However, in the form in which it appears in an excellent textbook for engineers[4] the same information is set out much more clearly:

> *. . . preventive maintenance should be considered when the following conditions apply.*
>
> *(1)   The time interval between equipment breakdown can be predicted with reasonable accuracy.*
>
> *(2)   The cost of preventive maintenance attention is less than the repair cost when both costs include that of any lost production.*
>
> *(3)   Equipment failure is likely to disrupt subsequent*

production operations or cause customer dissatisfaction.

(4)   Injury could result from equipment breakdown.

The list is easier to read and to remember than the same information written in a long paragraph. There are, incidentally, two basic forms of list: those in which individual items are simply 'bulleted' for identification, and those which are either numbered or lettered. On the whole, the former style is used when the order in which the points are read or dealt with is immaterial (a list of items of equipment might fall into this category), while the latter style emphasises the order in which the information is given (for instance, stages in a process). Lists are also sometimes numbered in order to allow cross-reference to specific items, or, as in the example above, to help the reader who needs to memorise the information.

☐ List information whenever it is possible to do so.

If a paragraph contains figures, measurements or dates, the planning stage includes checking exactly what information should be highlighted. Numbers might be better in tabulated form, or a comparison may be drawn by using 'parallel' sentences. In the following paragraph, there is an implicit comparison, but it is difficult to identify the details:

It is interesting to note that between January 1996 and March 1996, 1350 pcb were produced with nine operators, while between January and March 1997, seven operators produced 2869 pcb.

This can be rewritten more clearly as a mini-table:

January–March 1996: 9 operators produced 1350 pcb (150 per operator)
January-March 1997: 7 operators produced 2869 pcb (409.8 per operator)

The dates, number of operators and increased production figures are now easily identified and compared.

☐ Set out numbers in a way which is easy to use.

Layout is important both in the planning of paragraphs and in the organisation of whole documents. In the rest of this chapter, we look at some of the conventional formats which engineers use regularly in their working lives: memos and e-mail, letters and faxes, reports and sets of instructions.

☐ Use a conventional format with which the reader is familiar.

Use and Format

Memos and e-mail

Memoranda (memos) are in-company documents; they are still widely used but to a certain extent have been overtaken by e-mail, which is probably the fastest and simplest way for colleagues to contact one another. E-mail can be used on a worldwide basis, while the traditional memo did not travel outside the company, and was – and is – often circulated within a single building or one complex of buildings.

Traditional memos are essentially one-to-one correspondence, although copies may be sent to a small number of people apart from the main recipient, and they are occasionally put on a notice board for wider reference (for instance, a reminder about a general staff meeting). On the whole, memos do not travel far up or down the hierarchy of an organisation, but are sent between people of approximately equal status.

An e-mail message, on the other hand, may be sent to a large number of people at the same time, for instance by a lecturer to a whole class of students, provided, of course, that they are all on the e-mail system and can receive the message. This highlights one of the problems with e-mail: if the recipient does not choose to accept the information, it does not get through, and the writer may not be aware of this.

Although e-mail should travel safely from sender to recipient, this too does not always happen.

One of the widely claimed advantages of e-mail is that it does away with the need for a hard copy; in practice, if the result is to be used away from the computer (for instance, if it is the agenda for a meeting), it will have to be printed out. This will probably also happen if the recipient simply wants a permanent record of the message. In the past, memos were 'one-off' communications, which left no record behind them, but increasingly companies demand that memos are kept in some form, so that a copy can be made available for future reference. As a result of all this, e-mail sometimes ends up in a form which is indistinguishable from a memo!

Both e-mail and memos are less formal in style than other types of professional writing. Abbreviations such as *I'll* or *it's* are acceptable, although slang or jargon should be avoided unless the information is intended for one person only, who is well known to the writer. Spelling, punctuation and grammar should, of course, be correct, as in any business correspondence.

In this respect, e-mail tends to encourage a casual approach. It is perhaps too frequently used for social contact within an organisation, and 'junk mail' may tend to clutter up the system. As a result, writers get into the bad habit of producing informal notes rather than coherent sentences. Often, this will present no problem, but as we have discussed earlier, bad – or inadequate – punctuation or grammar can cause misunderstanding or ambiguity. For this reason, and for its comparative lack of confidentiality, e-mail should be used with discretion.

The printed memo usually has no more than two or three paragraphs, and points may be listed and numbered if this is helpful to the reader. It is always short, and deals with one subject only; a second page is very rare, and most memos contain fewer than 200 words. Memo envelopes, which can be re-addressed, are sometimes used; alternatively, a memo can simply be folded over and stapled. Because of this casual presentation, memos, like e-mail, are not suitable for the circulation of confidential information.

☐ Memos and e-mail consist of short, comparatively informal messages. They are not appropriate for confidential information.

The discussion of format and the example which follow both relate to memos, but there is a useful reminder there to the e-mail user. The headings will not be relevant – but writer and reader still need to be readily identified; a date is still important even if incorporated automatically, and the subject matter must still be made clear. If there is more than an odd line of text, it will need some simple structure, for instance listing, for the convenience of the reader. If the information is likely to be printed out, it must be usable in that form.

There is a conventional format for a memo, although the order of information may vary. Memos have four headings:

To   *Date*
From   *Subject*
                         (or Reference)

The names used may be formal (Mr J. Twigg) or informal (Jim Twigg), or may be replaced by a job title (Personnel Manager). The deciding factor is often the need to identify both writer and recipient without ambiguity. All business correspondence should of course be clearly dated, in a form which is universally recognised, that is, 4 June 1997 not 4/6/97, which could be understood, American-fashion, as 6 April 1997. The subject heading should be brief. It defines the material precisely, but should not itself be more than five or six words long. Sometimes a company reference number is used instead, or a relevant identifying number such as an order number.

As there is no introduction to a memo (that is, no equivalent of 'Dear Sir'), there is no conclusion (that is, no equivalent of 'Yours faithfully'). Memos are often initialled by the writer at the end, although some companies insist on a full signature.

The following memo was written by a tutor to engineering student 'project managers'; such a message is unlikely to be sent by e-mail, as the recipients were expected to keep the information available for reference. The essential information is included, but it is not easy to follow or to use. An improved version follows.

7/10/97

To all student 'project managers', Year 1.
From Jim Twigg, Year 1 Tutor.

I need to know the hazards which are associated with your project, remembering that if you take apparatus away from supervised areas special precautions are necessary. Let me know the steps you propose to deal with them. The following hazards may occur and should be guarded against: electric shock and insulation and protection, fire risks and where you intend to keep fire extinguishers and other fire-fighting equipment; if pressurised or other highly stressed components are involved they must be protected against; moving components must be guarded, such as rotating shafts, gears, belts, pulleys; you may also be injured by falling or tripping especially if there are sharp projections or objects are dropped. The first time hazardous apparatus is used, your project supervisor must be present and if it is very hazardous, every time. If you have problems with safety, I will assist you.

Comment

This memo has the virtue of being short, but it is not always sufficiently informative ('I need to know' – when?; what if 'I' am not available?). Its appearance is heavy and there are stylistic failings, as follows.

1. remembering: who remembers? Presumably the student, rather than 'I'.
2. them (line 4): this appears to refer to 'special precautions', but the sense would suggest that it refers to 'hazards', which is much earlier in the memo.
3. The list of hazards is written along the line rather than as a list; it includes 'where you intend to keep fire extinguishers', which is clearly not a hazard but a digression.

4. they must be protected against: against what? The sentence is the wrong way round.
5. The contrast of 'the first time' and 'every time' is badly placed, so that the significance of the latter is hidden.
6. No attempt has been made to give emphasis to particularly important information.

Improved version

To: All student 'Project managers/, Year 1.
From: Jim Twigg, Year 1 Tutor.
Date: 7 October 1997.
Subject: Year 1 Projects: Safety.

Please let me know before the end of October the hazards associated with your project and the steps you propose to take in order to safeguard against them.

In particular, you should be aware of the following hazards:

(1) *Electric shock*: insulation and protection must be adequate.
(2) *Fire*: notify me of the position of fire extinguishers and other fire-fighting equipment.
(3) *Pressurised or other highly stressed components*: protection in the case of failure must be adequate (see the appropriate Notes)
(4) *Moving components,* such as rotating shafts, gears, belts, pulleys: guards must be in place.
(5) *Sharp projections, dropped objects etc.* may cause injury from falling or tripping. Take care.

If apparatus is used away from supervised areas, special precautions must be taken; contact me in advance.

Your project supervisor must be present:
(1) whenever hazardous apparatus is used for the first time;

(2) *always* if the apparatus is rated 'very hazardous'.

If you have safety problems or queries, contact me on extension 274 (room 34B); if I am not available, you must speak to the Departmental Safety Officer **before** starting work.

Letters and faxes

The place of the formal business letter has in recent years been steadily eroded, both by greater use of the telephone and by the popularity of the fax. The former is outside the scope of this book (but see Further Reading, p. 129); the latter presents a particularly interesting challenge. Faxes are used essentially in two ways: firstly, to provide a rapid, informal, person-to-person exchange of information, particularly when the telephone would be inappropriate, for instance when diagrams or complex technical details have to be conveyed, and secondly, as a substitute for the business letter when rapid transmission is needed. This is particularly important when the recipient is in a part of the world which can be reached by post only with difficulty or after long delay. Few conventions have as yet been developed for the use of the fax, although some attempts have been made to suggest guidelines (see especially Notes[5]). Its informal use between individuals who are probably known to one another is much too widespread to be challenged, and yet there are problems: the fax is as binding as a letter; it may similarly represent the company which sends it; if it is handwritten, it may not be clear; it may be sent too rapidly to allow for second thoughts. This last is especially worrying: if the writer has time to think over the response, it will be considered, courteous and, one hopes, accurate. If it is sent immediately, it may be none of these things. There is no doubt that a faxed response can be of enormous value to both writer and reader – provided that the former is aware of the dangers.

The formal fax is used by many organisations in place of a letter. Clearly, it saves time, the cost of postage, and the trouble of ensuring that the sender catches the post. Reasonable confidentiality can now be built in (for instance, by the use of a password), and the fax can be linked directly to a computer for convenience. Yet writers are inclined to forget that if the fax is replacing a letter, it should have the safeguards of a letter: it should be carefully printed out, checked and, if appropriate, counter-signed. It should certainly be treated with the same respect.

The situation is helped if the company has produced a well laid out cover-sheet, to be used for all formal faxes. This will give the name and status or department of the sender, together with the company name and logo, address, telephone and fax numbers. The sender's direct line for telephone and fax should be added if appropriate. Similar information should be given for the recipient, not least to ensure that the fax reaches its intended destination as quickly as possible. The date and the number of pages should be given, and a short heading to highlight the subject. If all this is well designed, the fax can be as impressive as the company's business letters.

In style, the formal fax will be similar to any other business correspondence which goes outside its own organisation, although even this has changed subtly over the years. Letters have become much less formal than they used to be, and pompous letters (starting with *I am in receipt of yours of the 6th ult.* or some such horror) are a thing of the past (*almost!*). Letters have also become shorter, being used nowadays much less to convey a great deal of technical detail and much more as a brief introduction to, or over-view of, the subject (the covering letter). It is much more common for the detailed information to be presented separately as a report.

Nevertheless, there are still many occasions which demand a letter. Contact with individual staff, such as appointment, promotion, changes in working conditions or formal disciplinary action are in letter form. Letters may confirm the agreement reached (or not reached) in a meeting, or may send brief information requested by clients or customers. Engineers may have to write to members of the general public or to influential groups, for instance to ask for permissions, and of course any technical report is likely to be sent with a covering letter.

☐ Letters are always, and faxes are often, formal documents which represent the company that sends them.

In preparing to write any of these, or other, letters, the engineer needs to assemble useful information: technical data, previous letters on the same subject and up-to-date or revised details which must be included. It is helpful to look at the last letter received from the company to which the present letter is addressed, not only to make sure that all relevant topics are covered in the reply, but also to check the tone used. If the reader's company uses a formal style and formal names, it is generally wise to follow suit, and *vice versa;* it would suggest disagreement and perhaps hostility to write *Dear Mr Twigg* when he signed his letter *Jim.*

The use of first names is a tricky subject. Business letters once began *Dear Sir* or *Dear Sirs* as a matter of course. Although this is still not uncommon, surnames are now more widely used (*Dear Mr Twigg*) and first names are also used (*Dear Jim*). There is still need for caution. Jim, whom you know well, may be away from the office and your letter may instead be dealt with by Tom, whom you do not know. Worse, Jim may, no doubt by oversight, have failed to pay his bills, and it is difficult to write a stiff letter threatening legal action if you begin it, as usual, with *Dear Jim.* Discretion is needed, and previous correspondence is a useful guide to what is appropriate. Company policy, of course, may be the overriding factor.

In the case of a fax, the writer and recipient's names are already on the cover sheet, and writers are sometimes uneasy about how to begin. There is no reason why the *Dear Mr Twigg* (or *Dear Jim*) form should not be used, with the appropriate conclusion. Some writers prefer to begin simply with the name, especially if it is a first name (*Jim*). In this case, the conclusion can be *Regards,* although this is most frequently used between people who already know one another.

Engineers may be men or women. This seemingly obvious fact is sometimes overlooked by letter writers, and too many letters still start with *Dear Sir* when a little research would have shown that *Sir* is female. People are often unsure about the correct form of address to a woman, and indeed the situation is changing and the conventions are not entirely clear. *Dear Madam* is as appropriate as *Dear Sir*, although, oddly, it sounds rather more formal; *Dear Miss Twigg* or *Dear Mrs Twigg* is sometimes resented, and indeed a woman's married state, being irrelevant to the competence of her engineering, is often not shown. *Dear Ms Twigg* is generally acceptable, as is *Dear Jane Twigg*. It is helpful for women engineers to print their names under the signature in the form which they would like to receive in reply (another good reason for checking previous correspondence), and the full-name signature, as opposed to initial and surname, at least prevents the *Dear Mr Twigg* response. In passing, it is worth noting that an increasing number of women use their maiden names for professional purposes, thus adding to the confusion! It is worth taking trouble to find out other useful details about the recipient of a letter: first name or initial (needed on the envelope if not in the letter) and job title.

Nowadays, the conventional layout of a letter is usually along the lines of that shown in Figure 1.

The language in which a letter, or a formal fax, is written should be as simple as possible, unambiguous and courteous. If it is a reply, it is sensible to begin *Thank you for your letter of 6 July*. This is much better than *I have received your letter of 6 July* (how else could you be replying to it?) or *I am writing in reply to your letter* (obviously you're writing!), and more grammatical than *With reference to your letter of 6 July*, which is not a sentence. The recommended version also avoids 'I' as the first word of the letter, which can sound a bit self-important; whenever possible, start with 'you' rather than 'I'. For example,

> *I know that you are anxious that the completion date is as early as possible*

would be better as

> *You are naturally anxious that...*

Companies vary in their policy about *I*, *we* or the impersonal form (*It is agreed that...*). Such policy should be followed.

(a) ——————

(b) ——————

(c) ——————

(d) ——————

(e) ——————

(f) ——————

(g) ——————

(h) ——————

(i) ——————

(j) ——————

(k) ——————

Figure 1 Letter Format

(a) is the address of the writer, unless the letter is written on company headed paper. There is no punctuation within the address.

(b) is the date, with the month written as a word. The most common form of date is '7 May 1997'. If the balance of the page requires it, the date may be put at the left-hand side of the page at this level.

(c)   is the name and address of the recipient. This may take the form of status and address ('The General Manager, Bridgeco Engineering Co. Ltd'), although if possible the individual's name should also be used. Sometimes 'Dear Sir' or 'Dear Madam' may be used, with an extra line, 'For the attention of...', underneath. If a name is included, it should include an initial ('Ms J Twigg'). Again, there is no punctuation within the name or address.

(d)   is the greeting: 'Dear...' as appropriate. No initial is used at this point ('Dear Ms Twigg'), and the name is not followed by a comma.

(e)   is a heading, which identifies the subject of the letter. This may, for instance, be an order number or the name of a project or the same heading as in the letter replied to. The use of a heading enables the writer to begin the text directly, with the first factual information. It is also helpful to the reader, allowing the subject to be identified at once. Sometimes the heading is prefaced by *re:* (*with reference to*), but this is becoming old-fashioned.

(f)   represents the beginning of the text, which contains normal punctuation. Paragraphs are 'blocked' to the left-hand margin rather than indented, and double spacing is left between paragraphs.

(g)   is the 'courtesy close' sentence, which often causes the writer more difficulty than all the rest of the letter. It must be a grammatical sentence (see p. 24), which rules out *Looking forward to receiving your comments* or *Thanking you in anticipation*, which contain no verb. The phrase *in anticipation* used in this way should always be avoided: it is a particularly nasty piece of jargon. The nature of the letter will dictate an appropriate, short courtesy close, but *I look forward to hearing from you, Thank you for your co-operation* or *If I can be of further assistance, please contact me at the above address* are all regularly used and satisfactory.

(h)   is the conclusion: 'Yours...'. Convention dictates that *Dear Sir, Dear Madam* or any form of greeting which does not include the recipient's personal name will be followed by *Yours faithfully* (capital Y, small f, with no punctuation), to which there is no alternative. As soon as a personal name is involved (*Dear Mr Twigg*), the most common conclusion is 'Yours sincerely' (capital Y, small s) or, although it is perhaps seen as old-fashioned, *Yours truly.* Occasionally, if writer and reader are well known to one another, the conclusion can be expanded a little:

Kind regards.
Yours
Jim

This is acceptable as long as it is felt by both parties to be appropriate.

(i)   is the signature of the writer, or sometimes of someone signing on the writer's behalf.

(j)   is the typed name, including any relevant titles, such as *Dr*, which should never be added to the signature.

(k)   is the typed job title of the writer. These two pieces of information, (j) and (k), should ensure that the reply is correctly addressed.

When a choice is available, it is worth noting that 'we' (that is, the writer on behalf of the organisation) is easier to use than the impersonal; it is more concise and sounds more co-operative. *We agree that* gives a sense of personal concern which is missing from *It is agreed that. I* may be allowed, but should be reserved for personal involvement; when the writer means *I will be at the next meeting of the Institution and will be able to discuss the matter with you then*, the use of *we* would be silly. Letter writers are oddly unwilling to refer to themselves directly as 'me', and manage to sound very pompous in avoiding it. *The undersigned, the present writer* or even *myself* distance writer from reader; if writers mean *please contact me*, they should say so.

Above all, both letters and faxes should be helpful. They go as ambassadors of the writer's company, and if they are over-formal, ambiguous in content, wordy or chilling in tone, goodwill is strained. The writer should always be aware of the need to give clear information, offer assistance, sound concerned or offer apologies, as required, and to check that the document says exactly what it was meant to say.

Far too often, letters and formal faxes (and reports and other company documentation) are not checked. If they are dictated into a machine and then typed when the originator is far away, there is an especial danger that the wrong information is given. Somebody must be responsible for checking that the typed version is correct, and that it is attractively set out. If the writer is available, he or she should check the document carefully, for signing it is an acceptance of responsibility for any mistakes it may contain (see also Chapter 6).

The following information is a discussion of tests carried out on a swaged tube as part of a proposed jointing system. While the technical content is complete, the lack of organisation and paragraphing makes it difficult to read. If this information is to be conveyed to a client outside the writer's organisation, a decision about the most appropriate format has to be made. There is clearly too much detail for a telephone call, although the main result might initially be conveyed that way. A letter is a possibility, although again the amount of technical content would make it unwieldy; the same would apply to a fax in the same format. The most appropriate choice is a one page report with a short covering letter. Both documents might be sent by fax, although a copy would almost certainly follow by post. The information is set out below, and then given in the format recommended.

On 27 May, Richard MacNamara and I (Sarah Watson) performed a simple test on the 76 mm diameter swaged tube you supplied. We had made first a plaster cast of the inside of the tube and secondly a steel plug to match the cast. The plug was then forced through the tube held by a plate welded around the edge in a compression testing machine. The effect on the swaged end was then similar to that intended for the pull-out test being planned. The tube failed at a load of 218 kN by the cracking of the longitudinal weld (which had been ground flat on the inside previously) after very little distortion. This result would appear to confirm a difficulty with the jointing system being proposed, as suggested previously, and it is not clear what simple steps can be taken to overcome it. The failure load of 218 kN gives a factor of safety of only 1.3 over the normal working load of 168 kN in axial tension. To confirm this figure of 218 kN I suggest that several more specimens are tested in the same way, but it would seem prudent to give consideration to the weld strength problem before embarking on a more expensive series of pull-out tests. EN26 steel will be needed because it has been suggested that failure might require the swaging load of over 600 kN and because the fittings have been designed to be suitable for the thickest available tubes of 168 mm outside diameter (10 mm wall). However, the result of the preliminary test performed last week might indicate that a lower grade material would be suitable. Use of mild steel would save on the cost. As I have suggested, I think we should perform a few more simple push-out tests before going ahead with the pull-out ones.

Comment

The information in this passage is written in one long paragraph, with no attempt to identify the tests carried out, the results or the comments. It is difficult to decide what is fact and what is suggestion. Headings, sections and the use of space will make a great difference to the reader's comprehension. For these reasons, the most helpful format is probably a short report sent with a covering letter.

The letter is set out according to the conventions given on p. 77, signed and dated. The heading draws the reader's attention immediately to the subject, and the essential test results are given at once. There is, of course, some overlap with the report itself, but this enables the reader to take in the key message before he has time to consider the report as a whole. If he chooses to pass on the report to a colleague, he still has this message in the letter itself.

It has been assumed that writer and reader are known to each other, and the tone is therefore friendly but formal. The letter gives the writer the opportunity to add the kind of detail which would be inappropriate in the report – in this case, that she or a colleague will phone to discuss the action to be taken as a result of the report. The addition of *Best wishes* before the close of the letter reflects a good working relationship and, perhaps, the desire that it should continue.

The report is equally formal, and the use of the first person in the original passage has disappeared. In the case of such a short report, it might not be thought necessary to number the headings; with a longer report, they would certainly be numbered (see p. 86).

Both letter and report follow.

Improved version: covering letter

29 May 1997
Mr S Jenkins
General Manager
Meredith, Jenkins & Sons Limited
Station Road
Denfield
Yorkshire SH1 6DJ

Dear Mr Jenkins

re: 76-mm diameter swaged tube, as supplied

On 27 May, Richard MacNamara and I carried out a push-out test on the swaged tube, as agreed. Details of the test, the results and our comments are given in the enclosed report.

As you will see, the results we obtained confirm that there is a problem with the jointing system as proposed. We feel that more consideration should be given to the strength of the weld before the series of pull-out tests commences; as you know, these will be expensive. However, our preliminary work suggests that savings could be made by the use of mild steel instead of EN26, provided that strength is not compromised.

We would strongly recommend that further push-out tests be carried out as soon as practicable. Either Richard or I will ring you in a couple of days to discuss the best way for us to proceed.

Best wishes.

Yours sincerely

(signature)

Sarah Watson
Development Engineer

Report

Introduction
In accordance with instructions received from Mr Simon Jenkins of Meredith, Jenkins & Sons Limited, Richard MacNamara and Sarah Watson, Development Engineers, carried out a preliminary push-out test on the 76 mm diameter swaged tube supplied. The test was requested because of uncertainty about the suitability of the tube for the proposed jointing system.

The test was conducted on 27 May 1997 at the Research and Development workshop at Denfield.

Test procedure
A plaster cast of the inside of the tube was made, and a steel plug machined to match the cast. The plug was forced through the tube, which was held by a plate welded round the edge, in a compression testing machine.

The effect on the swaged tube was similar to that intended for the planned pull-out test.

Results
The tube failed at a load of 218 kN by the cracking of the longitudinal weld, which had previously been ground flat on the inside, after very little distortion.

Comments
This result appears to confirm a difficulty with the proposed jointing system, and it is not clear how it could easily be overcome. The failure load of 218 kN gives a factor of safety of only 1.3 over the normal working load of 168 kN in axial tension.

Recommendations
1. Further push-out tests should be carried out on other samples in order to confirm the figure of 218 kN.
2. Consideration should be given to the weld strength problem before a more expensive series of pull-out

tests is started. The components for the pull-out tests include EN26 steel, since failure might require a swaging load of over 600 kN, and the fittings have been designed for the thickest available tubes of 168 mm outside diameter (10 mm wall).

3.   Mild steel appears to be suitable for such tests, and would reduce the cost. Its use is therefore recommended.

Richard MacNamara
Sarah Watson

29 May 1997

Reports

Reports, like letters, are written for their readers. This obvious fact is overlooked at the report writer's peril. Most reports convey information about a given subject, such as a project, an accident, tests on equipment or a visit. Many reports are also requests: the writer wants more time, more money or more co-operation, to sell a product or an idea. Both formal language and logically presented format must work to the benefit of the reader, who is the most important person.[6] They will then also result in the 'right' response from the point of view of the writer.

The reader and the objective must both be clearly identified at the start of the report-writing process. This will help the writer to use the correct language, that is, the appropriate level of technicality, with explanation and back-up as required. Far too often reports, like letters, use pompous words (*initiate* rather than *start*) and inexact expressions (*in due course, regularly*). As in all technical writing, the English language should be used in a formal but flowing and readable way, and as precisely as possible. Technical language should be chosen for the reader and for the subject matter, preferably in that order.

The purpose of the report should also be well defined in the writer's mind, so that all the information given is relevant to the subject and builds up a clear, uncluttered picture of the situation and what should be done about it. This clarification also helps to make the writing concise; report readers are usually busy people who prefer a short document to a long one. There is a tendency among engineers to ramble, often repeating the same information several times, introducing unnecessary details or wandering away from the point.

Although some reports are read in order from start to finish, far more reports are *used*, with the reader picking and choosing sections which are helpful, of particular interest or needed urgently. The structure of the whole document must be apparent from the contents list, although convention is also a useful guide. A report on a technical process might follow a standard format such as the following:

Title page
Contents page
Summary   (including, very briefly, the main conclusions and recommendations)
Introduction   (with the reason for the investigation, and any constraints in carrying out the work)
Procedure   (how the investigation was carried out)
Results   (the facts which emerged from the investigation)
Discussion   (the implications of the results)
Conclusions   (what is or is not satisfactory)
Recommendations   (what should be done as a result of the conclusions)

A long advisory report might have even more sections:

Title page
Acknowledgements
Summary   (including the main conclusions and recommendations)
Contents list   (section numbers, headings, page numbers)

Introduction	(putting the reader in the picture, including any necessary background)
Findings	(what was discovered as a result of the work)
Conclusions	(the implications of what was discovered)
Recommendations	(what should be done in the future)
References	(books, articles etc. referred to in the text)
Bibliography	(other related reading)
Appendix or Appendices	(supplementary material)
Annexes	(other documents, bound at the end of the report for the convenience of the readers)

Diagrams will often be included, either in the main body of the report or at the end, as appropriate, although in this as in all decisions, the convenience of the reader must be considered. Generally, it is easier to follow the flow of information if the diagrams are placed near the words which explain them. Each diagram should be referred to in the text, with a number and a title to identify it. A more detailed discussion of the use of diagrams in reports can be found elsewhere.[5]

The format shown above is very complex, much too long for many reports, but the pattern is useful: introduction, factual content, comment or conclusions and, if they are asked for, recommendations. This ordering of information can be followed even if the report is a short one.

Within the given format, reports are closely structured in sections with numbered headings. A very short report might need only a small number of headings (see p. 81) and in this case there may be no advantage in numbering them. Most reports, and certainly any which are more than a couple of pages in length, have headings numbered for ease of identification and reference. There are various numbering systems, some of them very simple:

1. Background
2. Investigation
3. Comments

Longer reports need a more complex structure, either paragraph numbering or, preferably, decimal notation, in which all headings are numbered, irrespective of the number of paragraphs in the section. This is a clear hierarchical structure, in which sections with major headings can be subdivided into sub-sections with lesser headings, as follows:

1   MAIN HEADING
1.1   Subordinate Heading
1.1.1 Smaller heading

The pattern is repeated, so that sections 2, 3, 4 and so on are equal in importance to section 1, sections 1.2, 6.4, 9.5 and so on are equal in importance to section 1.1, sections 1.1.2, 1.2.6, 8.11.3 and so on are equal in importance to section 1.1.1. If the report is very long, a fourth layer, 1.1.1.1, can be added, but the numbering should not be extended beyond that.

This numbering system is useful in that it is widely used and recognised, easy to check and logical. Sections can be readily identified: section 2.4 begins with that number and continues to 2.5, and all material in between is subordinate to the 2.4 heading. In this system, all numbers have headings and *vice versa,* with one exception. Lists are kept clear of the main decimal notation system. Items are identified by single arabic numbers in parentheses to the left of the text, as in the example below:

Headings in reports should be:
(1)   short, usually not more than five or six words
(2)   as specific as possible
(3)   set out in a form appropriate to the numbering hierarchy.

This last point is particularly helpful to readers: if major headings look as if they are major headings (for instance, in bold upper case) and subordinate headings can be identified by their format (for instance, a smaller upper case bold, or lower case bold) as well as by their number, the reader can see the structure of any page of the document immediately. The format of the headings and the numbering system reinforce each other. All reports should be dated, not least to protect the writer, who is responsible for the accuracy of the information at the time of writing but who cannot be expected to foresee economic or legal changes which could invalidate the conclusions.

Ideally, a report is a document for *use,* planned and written for the convenience of the reader, and structured so that its logic is immediately apparent. The reader should be able to identify and use detailed information without having to work through long paragraphs or pages of text. For this to happen, all the headings must be set out in order on the contents page, together with details of any appendices or annexed material. Readers then have a logical overview of the structure of the document, and can select any section or subsection which they need, seeing how it fits into the report as a whole. They are also helped by the inclusion of a summary at the beginning of the report.

In some ways, the summary is the most important and influential part of the text. As has been suggested above (see p. 85), it has to be placed at the start, perhaps before the contents list and certainly before the introduction, because readers will turn to it before they begin to read any other part of the text. The summary gives a brief resumé of the whole report, emphasising those aspects which are likely to be of most importance to readers – almost always, the conclusions and recommendations. In about a third to half a page (the summary would be longer than this only if the report itself were long, perhaps more than thirty or forty pages) the writer has to give readers an accurate and unbiased picture of the message contained in the report. Readers will then read this first, go back to it as a reminder of the content, study it to help them understand technically complex information, and, in some cases, read the summary and nothing more. This is true especially of two categories of readers: managers who are not directly concerned with the work discussed but who need a general view of what is happening, and senior staff who are unlikely to give the time to a full reading of the report but who may well be involved in making decisions on the basis of its recommendations. If they read a clear, well-balanced summary, they are almost inevitably on the side of the writer. Needless to say, the summary is thus highly influential – and also difficult to write well. The time needed to produce a good summary should never be underestimated.

☐ Reports have a logical structure which should be clear to the reader; their language is always formal.

The importance of good structure and appropriate style to the success of a report is best shown in an example. The following information is given in unstructured 'essay' form, with paragraphs but without numbered headings. It is then given in structured report form.

The effect of acoustic ceiling tiles on noise at the County Swimming Pools, Northwich.
February 1997

After acoustic tiles had been installed in the ceiling of the Learner Pool at the County Swimming Pools, Northwich, Jim Downs, a member of the Parks and Leisure Sub-committee, was asked to liaise with Dr Andrew Poynter, Senior Lecturer in the Department of Mechanical Engineering at the local University of Abimouth, in order to investigate the effect of the tiles on noise levels.

Andrew Poynter had been involved in 1995 in testing noise levels at the Pool, after staff and parents had complained to the Council. At that time, he had recommended that the Council install acoustic tiles manufactured by Fraser & Macfarlane to an area of $114\,m^2$. He measured current reverberation times (see Table 1) and also noise levels. These latter were recorded as an average of $92\,dBA$, with a maximum of $103\,dBA$ and a minimum of $87\,dBA$.

The Council approved Andrew Poynter's recommendation, especially as both he and the tile manufacturer's representative predicted shorter reverberation times (see Table 1) and lower noise levels if the tiles were fitted. The new acoustic ceiling was therefore installed in November, 1996.

Two months later, Dr Poynter was asked to assess the effect of the ceiling, and to give Jim Downs a report on the findings. He agreed to do this. On February 1, 1997, they visited the Learner Pool together when there was the usual weekend attendance of about 35 people, adults and children. This was approximately the same number as in 1995 when the original tests were carried out.

Andrew Poynter found that an area of $180\,m^2$ had in fact been treated with the acoustic tiles, but he did not consider that this invalidated his original predictions, although of course the figures were no longer precise.

The backing of the tiles was fibre glass in film bags, which was also different from that used in the earlier tests. This backing could have been responsible for the beneficial results which were obtained.

The reverberation times were measured and compared with the times predicted by Andrew Poynter and by the representative of Fraser & Macfarlane. They were found to be much shorter (see Table 1). Noise levels were also tested, with the average now 77 dBA, the maximum being 87 dBA and the minimum 71 dBA. This was a pleasing decrease over the earlier level.

Dr Poynter also talked to the staff at the Learner Pool. The instructors and the pool attendants all agreed that the general noise level had gone down, which made their working conditions pleasanter. They also felt that the intelligibility of speech had been greatly improved, so that they could now both hear more clearly and identify who was making a particular noise. This was obviously good from a safety point of view, and made the training of youngsters easier. They declared themselves to be delighted with the new ceiling.

After he had completed his assessment, Andrew Poynter reported that he was so pleased with the results that he felt the Council should extend the acoustic ceiling. The general noise level had clearly fallen and the reverberation times were much shorter, probably as a result of the backing of the tiles. The same kind of tiles should obviously be used in the future. The staff felt that their work had become easier and the safety level had improved, which must be good for everyone.

Andrew Poynter prepared his report along these lines and presented it to Jim Downs for consideration by the Council.

Table 1 below shows the measurements of reverberation times.

TABLE 1   *Reverberation times in the Learner Pool (curtain open)*

Frequency (Hz)	1995 (seconds)	A. Poynter's recommend- ation (seconds)	Fraser & Macfarlane's prediction* (seconds)	1997 (seconds)
125	1.75	–	2.8	0.7
250	2.2	–	2.0	0.4
500	2.4	1.5	1.3	0.55
1000	3.8	1.65	1.5	0.8
2000	3.6	1.65	1.7	0.85
4000	3.0	1.5	1.6	0.9
8000	–	–	–	0.55

* Fraser & Macfarlane's prediction was based upon $114\,m^2$ of treatment whereas the actual area installed was $180\,m^2$.

Comment

This is not a report, but an essay. It has no headings and lacks any clear structure. One set of measurements is given as a table, while the other is hidden within the text, and it is not easy to pick out background information as opposed to the current test. Sections of information cannot be quickly identified, and the reader would find it difficult to use the table at the end when reading the main text. Conclusions and recommendations are confused, and the 'report' seems to peter out with no clear statement of what action should be taken. There is no summary to give the reader an overall view of the essential message of the report; as a basis for discussion and action, it leaves much to be desired.

Improved Version

The information can be organised into a simple report format with a summary and a clear logical structure, as follows.

The effect of acoustic ceiling tiles on noise at the County Swimming Pools, Northwich
February 1997

Summary
In response to a request from the Council, Dr Andrew Poynter assessed the effectiveness of the acoustic tiles which had been fitted in 1996 to the ceiling of the Learner Pool at the County Swimming Pool, Northwich. Reverberation times and noise levels were found to have fallen significantly, giving additional directional information and speech intelligibility. Safety had been increased and staff working conditions had improved.

It is recommended that the Council should consider using similar tiles to extend the area of the acoustic ceiling.

1. Background
During 1995, a problem with noise levels in the Learner Pool at the County Swimming Pools, Northwich, was identified by Northwich Borough Council. Dr Andrew Poynter, Senior Lecturer in Engineering at the University of Abimouth, carried out a series of tests to ascertain both the reverberation times at different frequencies and the noise level, when an average weekend attendance of 35 people was present. The results of these tests were given in the report and are reproduced in Tables 1 and 2.

As a result of these tests, it was recommended that acoustic tiles manufactured by Fraser & Macfarlane should be fitted to $114\,m^2$ of the ceiling area, with the expectation that there would be an improvement in both reverberation times and noise levels. The predictions of both Andrew Poynter and the manufacturer's representative are also given in Tables 1 and 2.

The tiles were fitted in November 1996. In January 1997 the Council asked Andrew Poynter to assess the effect of the new ceiling.

2. Assessment of acoustic ceiling
On February 1, 1997, Andrew Poynter inspected the new

acoustic ceiling. As on the previous occasion, there were approximately 35 people using the pool.

The investigation showed that the tiles had been fitted to an area measuring $180\,m^2$ instead of $114\,m^2$. The backing of the tiles, fibre glass in film bags, was not precisely the same as that tested earlier. The predictions were therefore not accurate, but gave a sufficient indication of the changes which might have been expected. Reverberation times and noise levels were again measured.

2.1 *Reverberation measurement*

The reverberation times were measured and compared with those of the untreated ceiling and with the predictions made by Dr Poynter and by a representative of the tile manufacturers. The results are given in Table 1.

TABLE 1 *Reverberation times in the Learner Pool (curtain open)*

Frequency (Hz)	1995 (seconds)	A. Poynter's recommendation (seconds)	Fraser & Macfarlane's prediction* (seconds)	1997 (seconds)
125	1.75	–	2.8	0.7
250	2.2	–	2.0	0.4
500	2.4	1.5	1.3	0.55
1000	3.8	1.65	1.5	0.8
2000	3.6	1.65	1.7	0.85
4000	3.0	1.5	1.6	0.9
8000	–	–	–	0.55

* Fraser & Macfarlane's prediction was based upon $114\,m^2$ of treatment whereas the actual area installed was $180\,m^2$.

2.2 *Noise level measurements*

The noise levels were then measured, and compared with those recorded for the untreated ceiling. The results are given in Table 2.

TABLE 2   *Noise levels*

	Untreated ceiling (dBA)	Acoustic ceiling (dBA)
Average	92	77
Maximum	103	87
Minimum	87	71

2.3   *Discussions with staff*

The effect of the new ceiling tiles was discussed with attendants and instructors at the Learner Pool. They reported that the intelligibility of speech had greatly improved, increasing both the response to training and therefore safety. The overall sound level had been reduced, which was beneficial to staff, and they were better able to identify the direction of noise, which again was a safety factor. Overall, they were delighted with the new ceiling.

3.   Conclusions

As a result of the investigations described, Andrew Poynter came to the following conclusions.

(1)   Reverberation times were shorter than expected, perhaps because of the structure of the backing of the tiles.

(2)   Noise levels had fallen considerably.

(3)   As a result of (2), safety has been improved both by an increase in directional information and by improvement in speech intelligibility. The work of attendants and instructors has been made easier and pleasanter.

4.   Recommendations

In the light of these conclusions, it is recommended that the Council should consider extending the area of acoustic tiles, and that, if this were to be done, the same type of tiles should be used.

Andrew Poynter, BSc, PhD, C.Eng
Department of Mechanical Engineering
University of Abimouth
February 9, 1997

Instructions

Instructions are given in various forms, which have their own conventions and appropriate style. Specifications are perhaps halfway between reports and procedures: they present the standard expected or against which the work can be tested, and for this reason often incorporate or are closely linked to British Standards. Specifications may show, for example, how machinery is to be manufactured or maintained, or how a system is to be designed or operated, and the language in which they are written must conform to specific rules. Procedures are of two sorts: general procedures, which indicate that the work should be done in a particular way, and specific procedures, which give precise instructions for carrying it out. Instructions themselves simply tell the user what to do, often with little elaboration – although sometimes explanatory notes or diagrams are included.

In all these forms of instruction writing, a major source of difficulty is the use of a group of words which are often confused: *can/could, may/might, should/would, will/shall/ must.* They are defined as follows.

Can/could are used to show ability:
   The car can reach 110 mph.
   It could travel even faster in different road conditions.

May/might show permission or possibility:
   The car may (is allowed to) travel at 70 mph on the motorway.
   The lorry might arrive (there is the possibility that it will) before dark if it is not held up by a traffic jam.

Might can also be used negatively:
   The lorry might have arrived if the road had not been flooded.

Able means skilled or equipped to perform a specified job; it has no suggestion of willingness or intention:
   The engineer is able to repair the damage. (He or she is capable of repairing the damage.)

Should/would are nowadays used interchangeably, and their force is often ambiguous, as in the following examples:

> *I wonder if I should drive so fast.*
> (*should* implies hesitation, whether I ought to or not)

> *I should go as quickly as possible.*
> (*should* implies moral imperative, under an obligation to)

> *He would go if conditions were right.*
> (*would* implies unlikely possibility, conditions are not right)

> *He would go in spite of our warnings.*
> (*would* implies determination, he insisted on going)

As a result of this confusion, the choice between *should* and *would* is often one of sound, as at the beginning of letters:

> *I should be grateful if you would kindly supply the following.*

Will is also ambiguous in writing, although in speech it is often made clear by emphasis, as in the following examples:

> *I will mend the car tomorrow.*
> (*will* implies future action)

> *I will mend the car tomorrow in spite of my other commitments.*
> (*will* implies determination)

> *The garage will be responsible for those repairs which are covered by the guarantee.*
> (*will* implies future obligation)

All the above examples suggest the difficulty of using should, would, or will casually in technical writing.

On the other hand, *shall* is often used to express obligation, and is now so forceful that, particularly in specifications, it has become an instruction word:

The company shall be liable for the cost of maintaining the equipment.
The engineer shall carry out the repairs as agreed.

In the writing of specifications, 'shall' should be used as the normal word to represent obligation: the company is deemed to be liable, the engineer will have to carry out the repairs. In other contexts, there is not the same degree of uniformity, and the writer should remember that 'shall' can be interpreted as a simple future:

I shall go to enquire about the order tomorrow or the following day.

and must make it clear if the word is used to convey an obligation.

The other word which is available to writers of specifications is *must,* which allows no option, and which carries an extra layer of meaning: it tells the user that the obligation is not simply that of the current document. The action is mandatory because of 'higher' authority, for instance because the law demands it. This distinction between *shall* and *must* is a useful one, and should be maintained, not least because it is widely recognised.

For this reason, it is better in specification writing to avoid less precise expressions such as *is to/are to/has to*. In everyday life, we use them to suggest obligation, but in a technical document they can cause uncertainty: if this action *shall* be done and that one *has to* be done, are there different levels of obligation, and if so, which is the stronger? There is, of course, no recognised answer to this.

Most companies nowadays have written procedures which are regularly appraised and updated, and here too there is a recognised terminology. A general procedure describes how actions should be taken (it does not give instructions) and the word *should* is used, exactly as it was earlier in this sentence. Specific procedures, however, are much more like work instructions, and may be written either in specification terms (*shall* and *must*) or in the usual language of direct instructions, using the imperative form of the verb. This is the 'command' form, 'do this', as in the following examples:

Check the equipment before signing the form.
Inform the safety officer immediately if the machine is to be moved.
Replace the guard on the machine immediately after maintenance.

In these sentences, the verbs *check, inform* and *replace* have the force of a command; there is no possibility of misunderstanding the intention.

The engineer, in preparing instructions, is not in the business of asking, suggesting, recommending or preferring. Orders have to be carried out exactly, and the only safe way of writing is to make each step sound like an order. We are not asked if we would mind not smoking as we take petrol at the service station: the results would be too awful if we decided not to comply. Nevertheless, it is sensible to give a reason for the order unless it is as obvious as the *no smoking* instruction. *No unauthorised entry* may be followed by the familiar radioactivity symbol to warn us of the hazard.

British and International Standards combine to make many hazard warnings international in use, and the colour codes (red for prohibition, yellow for caution, blue for mandatory actions and green for safe conditions) should, of course, be adhered to. Words and symbols are both used in warning notices, and a combination is often sensible (for example, the *no smoking* sign is often accompanied by a crossed-out cigarette).

Instructions which are also warnings should be written in a positive rather than a negative form if at all possible:

Do not leave equipment switched on.

is better as:

Switch off equipment.

Some instructions, however, are negative in themselves, even doubly negative, as in:

Unauthorised persons may not change grinding wheels.

Both style and layout are important considerations in the preparation of instructions, as they are in the writing of reports. It is easy to see why. Instructions have to be carried out; they must be unambiguous or the wrong action will be taken, and they must be well set out or actions may be omitted or performed in the wrong order.

Correct choice of words is essential throughout the document. The writer must select words not only for their accuracy but also for the reader's understanding. Instructions are usually written for people junior to the writer in status, knowledge and experience: the difficulty is to think oneself back into the position of the reader, and to write accordingly.

Imagination is needed to re-create the reader's lack of knowledge and also to envisage the reader's physical position. Words such as 'left', 'right', 'back' and 'front' depend on point of view. Similarly, words such as 'near', 'far', 'close', 'beside', 'behind', and even 'big' and 'small', are subjective. Precise information is needed. Other words create problems: appropriate, relative, substantial and suitable all depend on the reader's perception, and should be clarified or avoided.

The organisation and layout of instructions are also important. The most important rule is: one step at a time in the logical order. Each step should be numbered, and sequential arabic numbers are the easiest to follow. Numbering identifies clearly the individual stages to be completed, and is a useful reference tool if the process is interrupted (each step can be ticked off as it is finished).

The writer is also helped by a numbered sequence as it helps in checking that all stages have been included in the correct order, that is, the order in which the reader will carry them out. Never back-track in instruction writing: imagine being told that the battery must be checked after all naked flames have been extinguished. The *first* action (extinguishing all naked flames) must come before the *second* action (checking the battery).

Space is an important tool, to limit the amount of information which the reader has to take in at one time. Space should be left between heading(s) and instructions, and each instruction should begin on a new line. General comments, such as *This procedure should be carried out weekly*, should be either at the beginning or at the end, but clearly separated from the instructions themselves. If other additional information is needed within the text, it should be distinguished either by position (for instance, in a margin to the right of the instructions) or by font (for instance, by being in italics). Warnings must be especially clear, and if possible placed both at the beginning and at the appropriate stage of the instructions. Attention should be drawn to them, by the use of red and (not least because colour-blindness is common) capital letters, underlining or some other identifying mark.

Sometimes instructions can be grouped, with extra space at the end of each sequence, and with appropriate headings. Whatever form is chosen, the pattern of instructions should be clear to users, as it will increase efficiency and add to their confidence that they are doing the job correctly.

☐ Instructions must be appropriate in language and set out in a way which helps the user.

The following set of instructions for the design of a state machine is written in a rambling and imprecise way. An improved version is given later.

First of all, you should make a statement defining the state machine in terms of a state diagram and then after the number of required state variables has been determined and the state representations chosen, you can determine both the next state functions of the present state and inputs and the output functions of the present state and inputs.

Comment

The instructions given above are written as continuous prose and are therefore difficult to use. The particular problems which an engineer would face in trying to follow the information are given below.

1. The instructions are given as one fifty-six word sentence.
2. 'First of all' is wordy; stages should be given numbers.
3. Users are told that they 'should make a statement defining...'. Does this mean that they ought to, but do not have to? 'Make a statement' adds nothing to the sense.
4. 'and then after...' Does this means that users could *start* by determining the number of required state variables? Are they being given the second stage, or not?
5. 'has been determined'. Users were told that *they* 'should define...'. Now some unidentified person is going to 'determine'. Is this a separate function of somebody else? The passive is always unhelpful in instructions, as it leaves too many questions unanswered (see also p. 106).
6. 'chosen': another passive verb. Who chose, and when? Are 'determined' and 'chosen' two stages or one?
7. 'you can determine' well, yes, but *should* they? *Must* they?
8. 'you': who is 'you'?
9. 'both': there are two sets of functions to be determined. Both at once?
10. 'and', three times. How should this section be divided?

The instructions can be rewritten with all this confusion removed, as follows:

Design of a state machine
(1) Define the state machine in terms of a state diagram.
(2) Determine the number of required state variables.
(3) Choose state representations.
(4) Determine the next state functions of the present state and inputs.

(5) *Determine the output functions of the present state and inputs.*

Key Ideas

- Paragraphs have unity of theme (p. 65).

- Good paragraphing produces space on the page and encourages the reader (p. 65).

- List information whenever it is possible to do so (p. 67).

- Set out numbers in a way which is easy to use (p. 68).

- Use a conventional format with which the reader is familiar (p. 68).

- Memos and e-mail consist of short, comparatively informal messages. They are not appropriate for confidential information (p. 70).

- Letters are always, and faxes are often, formal documents which represent the company that sends them (p. 75).

- Reports have a logical structure which should be clear to the reader; their language is always formal (p. 89).

- Instructions must be appropriate in language and set out in a way which helps the user (p. 100).

---

5 Good Style

Definition of good style ■ readership and objectives ■
accuracy ■ brevity ■ clarity ■ formality ■ active or
passive ■ signs, symbols and abbreviations ■ numbers ■
diagrams ■ examples ■ references ■ abstractions ■
'link' words and phrases ■ critical listening

Good style can roughly be defined as style which is appropriate to the needs of the reader. A particular piece of information might, for example, be presented as original research in the publication following an academic conference. Subsequently, it might appear in a textbook for undergraduates; it might become such a basic contribution to knowledge of the subject that it would appear in a newspaper article for the intelligent but non- specialist reader; it might be published in school textbooks; eventually, it might be found in a children's encyclopaedia. In each manifestation, the information will be presented with vocabulary, sentence structure, explanation and examples suited to the current readership. If the level is estimated incorrectly, the information will not be accepted. It will be seen as 'too difficult', 'bewildering' or 'condescending', and will be rejected by the reader for whom it was intended.

Such extreme variations of readership are unusual, but they illustrate the need to write in a style which is helpful and encouraging to the reader. Writing in a vacuum, for the sake of writing rather than for a clearly defined audience, is unlikely to result in good style, and, indeed, the end product is unlikely to be read at all. The first requirement of good writing is that it suits the reader.

☐   Good style is appropriate to the needs of the reader.

Earlier in this book, we considered choice of words, organisation of sentences and paragraphs, and appropriate format. In doing so, considerations of style were important. Sentence length, for example, can help or hinder reading: very long sentences are nearly always difficult to read. They may have their place in works of literature, but not in the immediate transmission of engineering information. However, good style is more than correct choice of words or appropriate length of sentences. It also involves a wise choice of material and the appropriate putting together of discrete pieces of information. Both aspects will be considered in this chapter.

Long before the first word of a technical paper or report is written, the prospective writer needs to ask and answer a series of questions:

Who is my reader/are my readers?
How much does the reader know about this subject?
Why does the reader want to read this document?
What do I want to tell my readers?
What result do I want from their reading of my document?

The reader or readers must first be identified so that the writer can choose the style correctly. How technical should the vocabulary be? How many terms will need explanation and how much can the writer take for granted? What kind of examples will be helpful?

When the answers to these first questions are clear, the writer can move on to objectives – his or her own, in writing the document, and the reader's, in sparing the time to read it. Why should a busy writer find it necessary to commit this information to a written form? Why should the reader, busily involved with other work, bother to read this document? One clear result of asking these questions is that the writer will recognise the need for brevity. Both writer and reader are busy engineers with other preoccupations than the paper in question, and they will be helped if it is as short as possible. Part of the courtesy of good style is to avoid wasting the reader's time with what is unnecessary or irrelevant.

These final questions, about objectives, need more careful analysis. What does the writer hope will happen as a result of this writing? The answer may seem obvious: she wants her new young assistant to be able to carry out instructions competently and in safety; he wants the answer to his letter to be an order for his company's product; they (much technical writing is the output of a group rather than of an individual) want their report to result in the company installing the most appropriate system. There are, however, other, hidden, motives: to persuade the client to think well of the company, to encourage superiors to think well of the writer; to further a promising career.

If the writer is to be successful in all these objectives, the information must not only be presented briefly. It must be accurate and it must be clear to the reader. Chapter 6 will look at accuracy of presentation, but good style involves choosing words and examples which convey precisely what the writer means, and which will be understood by the reader in the same way. Ambiguity or careless use of words will blur the issue and may produce the wrong result. Good writing demands precision of thought and precision of language.

☐ Good style has its ABC: accuracy, brevity, clarity.

The 'courtesy' of good style was mentioned above. Courtesy involves consideration for the reader, which may in turn dictate the way in which he or she is addressed. *He or she* is used advisedly; too many engineering documents assume that the readership is male. In letters especially, the address should be correct (see p. 75 for a more detailed discussion of this point), but in longer documents, the writer may feel that a decision has to be made. *He or she* used repeatedly is cumbersome and long-winded; the decision made in writing this book was to avoid prejudiced language (often by the use of the plural), and to keep the use of *he* or *she* to a minimum. It is of course never acceptable to generalise *he* for the engineer and *she* for the secretary, and writers should always be aware of the possibility of giving offence.

Courtesy will also be a factor in decisions about formality. On the whole, memos and e-mail are informal, faxes may be formal or informal (see p. 73),letters are more formal, and reports and technical papers are very formal, although there are exceptions in each case. Instructions are impersonal (not *You must switch off the engine* but *Switch off the engine*). Formal writing does not allow the use of abbreviations such as 'it's' or 'can't', or the inclusion of slang or casual expressions better suited to the spoken language. In the most formal writing, the reader is not addressed directly (*You will find that* becomes *It will be found that*), and abbreviations such as *e.g.* or *i.e.* are often written out in full, especially in passages of continuous prose. Sometimes company policy dictates the level of formality, but if the engineer has to make the decision, it will be based on a consideration of the company's relationship to the reader and also on the information itself. If a full report format is chosen, the style must be formal, although a short report (little more than a memo) for the use of colleagues might be written in a slightly less formal style. Once the appropriate style is chosen, it must be used consistently; any deviation to a more formal or less formal style will distract the reader and interrupt the flow of the passage.

Formality may also dictate whether the writing is active or passive. An active style moves directly from subject to verb to object:

I recommend this policy.

The use of 'I' may be considered too informal for company style, or the writer may want to stress that he or she is writing as part of the company, in which case a more suitable form might be:

We recommend this policy.

However, courtesy or company policy may dictate that the passive is used. The object (policy, in the example given) now becomes the subject, and the person carrying out the action is not mentioned, unless in the cumbersome and unlikely form *by me*. The sentence will now read:

This policy is recommended.

The personal 'I' or 'we' has disappeared, and it is no longer clear who has made the recommendation.

Traditionally, formal scientific and technical writing has used the passive, but recently this has tended to lose favour, as part of a general move towards greater informality in writing. It is inclined to be long-winded, as in the following example:

It is recommended that the new staffing levels be applied by each department as soon as possible.

This sentence contains seventeen words, while the active form:

We recommend that each department applies the new staffing levels as soon as possible.

contains only fourteen. The passive also tends to detract from individual responsibility:

I recommend...

is a personal opinion, while

We recommend...

is agreed by the company, but

It is recommended...

remains unattributable.

There is one by-product of such use of the passive which is worth noting. In changing from an active statement to a passive statement, it is easy to give universal significance to what was intended in a more limited way:

I believe that...

introduces my own opinion, while

It is believed that...

suggests that many other people agree with my belief. Indeed, when

I cannot accept the idea that...

becomes

The idea is unacceptable that...

the world is clearly on our side.

Modern style, then, tends to prefer the active to the passive, but there can be no absolute rule, good style being essentially style which is appropriate to the occasion and to the reader.

☐ Use the active rather than the passive, whenever possible.

Courtesy to the reader also demands that help be given over those aspects of the document which might cause confusion. Signs, symbols and abbreviations are all potentially ambiguous, and the writer should again ask a series of questions:

Will I use technical (or other) signs, symbols and abbreviations in my document?
Can I reasonably expect that the reader will interpret them correctly?
Is there guidance which I should give, for instance, should I quote an appropriate British Standard?
Would a glossary page be helpful to my readers?

Guidance is available to the writer (see Further Reading, p. 130), and this should in turn be given to the reader. As far as abbreviations are concerned, there is a traditional form of explanation (at the first usage, the term in full followed by the abbreviation in brackets, and thereafter the abbreviation alone). This is appropriate for a technical article, but is less useful in a report, which may not be read in order, beginning to end. An explanation in the Introduction, or on a separate glossary page, allows the reader to check a meaning from any point in the text. However, if the term is used only two or three times in a lengthy document, it is often easier to write it in full each time rather than to make the reader look for an explanation.

Measurements and numbering also follow conventions, and it is wise for an engineer to follow British Standards or SI Units, or guidance provided by one of the Engineering Institutions, whenever possible. There are sometimes areas of possible confusion caused by common usage, such as m, which is the abbreviation for metre or metres, but not, as it is widely used, for miles. Arabic numerals should always be used in preference to roman numerals, which can be confused and which in any case are not so easily recognised or understood nowadays. Numbers should also be given in three-digit groups (674 320, with a space rather than the traditional comma to separate thousands from hundreds), and a full stop is the most easily recognisable form of decimal point (as opposed to the comma, widely favoured in the rest of Europe and officially accepted in the UK as an alternative decimal point).

Small numbers, up to one hundred, are usually written as words, while larger numbers are given as figures. This convention should be applied with care, as there are various exceptions. Round numbers are always given in words (*about a million miles*). If many large and small numbers are quoted in the same text, it may look better to use figures in every case, including the occasional small number. No sentence should ever begin with a figure: write the number as a word, or, if necessary, change the word order in the sentence. Decisions about the presentation of numbers are best made in the light of the context as a whole, with the convenience of the reader uppermost in the mind of the writer.

□ Follow national, international or professional conventions in presenting signs, symbols, units of measurement and abbreviations.

Some numerical information is most usefully given in diagrammatic form. The various types of diagram available are outside the scope of this book (see Further Reading, p. 130), and in any case different forms of diagram can be tried out and compared with the aid of a good computer graphics package. However, there are still decisions to be made by the writer. Good style dictates that a diagram must not interrupt the flow of the reading. It should not divide a sentence in two, or disturb a closely reasoned argument. Although all the diagrams can be gathered at the end of a document, this can be irritating to the reader who needs immediate access to a diagram while studying the text. More use is likely to be made of diagrams which are placed, if possible, where the reader needs to see them, that is, interspersed with the writing; they should also be sensibly positioned on the page, for instance not too far from the textual reference, and with adequate space left on all sides. 'Crowded' diagrams are not easy to use. A diagram should ideally be integrated into its text, introduced, presented and then discussed, so that the reader is led through the section in a logical manner. If diagrams are purely supplementary and not immediately necessary to the understanding of the documents, they can, of course, be placed in an appendix at the end.

☐ Diagrams should elucidate and not interrupt the message.

Diagrams are intended to clarify the text, by showing trends, or making comparisons, or giving detailed scientific data. The writer must decide how much information it is appropriate to give: sometimes much more detail is available than readers are likely to need, and on other occasions only a few readers will want all the figures. In the latter case, the diagram in the text might be very simple, while a table in an appendix could give detailed figures for those who need them. As in all aspects of good writing, the convenience of the reader is the paramount consideration.

Sometimes the impact of the text is enhanced by the use of an example or a simple analogy which aids the understanding of a difficult idea. There are two rules for using examples:

they must be at the correct level for the readership, and they must be widely understood. The writer should not 'over-explain', giving examples when the point is already sufficiently clear. At the same time, an analogy which is woven into the text, as in the following example from the essay of a student engineer, can be helpful to the understanding and also enliven the text, adding a light-hearted touch to a serious subject.

The problem of progressive collapse can be likened to the toppling of dominoes. As one domino becomes unstable and falls, the impulse that it implants on its neighbour will cause further destabilisation. This will continue until the entire line of dominoes has collapsed. Stronge (1986) has evaluated the conditions required for the propagation of this wave of destabilisation. Obviously, if the separation, l, of the dominoes is greater than the height, L, the destabilisation of one domino cannot have an effect on the next. Less obvious is the fact that if the dominoes are closely spaced, the wave of collapse will run only for a limited distance before stopping. The spread of the collapse wave is fuelled by the release of potential energy stored in the upright dominoes. If this energy were not available, the process would be arrested by friction. In structures, failures may progress catastrophically, as in the case of Ronan Point, where a single failure spread, domino fashion, throughout the structure.

In this passage, the writer has extended his example beyond the obvious, and has then developed his principal discussion, of progressive collapse in buildings, on the basis previously established, of the collapse of the dominoes. There is one potential problem. Pleasant though the example is, it will work only if the reader is familiar with dominoes. A writer must always be alert to the danger of examples which seem clear in one cultural context, but which are not helpful to the rest of the world.

☐ A well-chosen example helps the reader's understanding and gives life to the text.

The passage above illustrates a common form of textual reference to books or articles quoted, that is, the author's name and the date of publication. An alternative form is a small superscript number used for each reference; the number is sometimes in brackets. Either style is acceptable, although the latter is better avoided in papers which contain mathematical material, as the superscript number is easily confused with, for instance, such numbers in equations. Whichever form of textual mark is chosen, full bibliographical details must always be provided at the end of the document. These include, for books, the name of the author or authors, the title, publisher, place of publication, edition number (unless it is the first edition) and date, as in the example below:

> Fisher, Barry: *How to document ISO 9000 Quality Systems*, Ramsbury Books, Marlborough, 1995

The appropriate details for articles are author's name, title of article, name of journal, volume number, date and page reference, as follows:

> Ward, Anthony: Measuring the product innovation process, *Engineering Management*, volume 6, number 5, October 1996, pp. 242–6

The exact format for presenting such information varies, and it is often wise to consult the major journals in the writer's own discipline and to follow the style which they use. Needless to say, the essential qualities of all references are that they should be accurate and consistent.

☐ References should show full and accurate bibliographical details, and should be consistent in form

Engineers who follow all the rules and guidelines for good writing sometimes continue to feel that their documents do not flow well. Good style includes a felicity in choice of words and a sense of the rhythm of a passage which come naturally to some writers (of all disciplines) but which are not easily achieved by others. Nevertheless, it is important to be aware of some of the ways in which writing style can be improved, even if a high standard of literature is beyond the reach of the writer. Reading one's own writing out loud and listening critically is a good way to improve style, as awkward repetition, poor sentence structure or abrupt paragraph endings often become obvious as we hear them. For instance, the sentence before last originally ended with *beyond the writer's reach*, but the repeated 'r' sound is ugly, and a simple change to *beyond the reach of the writer* immediately improved the style. Two more real-life examples from the writing of engineers show how improvements in style can be made.

> *The problem was exacerbated by the existence of unclear perceptions by those engineers involved in the project of who the client really was.*

If this sentence is read aloud, the awkwardness of the repeated 'by... of' construction is apparent. There is a secondary repetition of sounds in 'exacerbated' and 'existence', but a much more serious problem results from the use of two heavy abstract words, 'existence' and 'perceptions'. The sentence also ends in a weak style, with a run of short words following the long and complicated words – there is almost a sense of anticlimax – and the meaningless word 'really'.

The first problem to tackle is that of the two abstract words. 'Existence' simply means that something *is*, and 'perceptions' means what they (the engineers) *saw* or *understood*, or, in this case, failed to understand. Once this part of the sentence is simplified (and the long word 'exacerbated' replaced by the simpler 'made worse'), the repetitions will disappear. Out of the mass of words, a simple statement appears:

> *The engineers involved in the project were unsure of the client's identity. This made the problem worse.*

☐ Avoid abstract words as far as possible. Use simple, direct language.

*At present no fax is logged, they used to be but they got too
busy to continue the system.*

The first part of this sentence, 'at present no fax is logged', is
clear enough. The second part (which is in any case grammatically a separate sentence) begins with 'they', but as this
word follows no plural noun, we do not know who 'they' are.
'They' also 'got too busy', which presents us with a second
unidentified 'they'. Do both 'theys' refer to the same plural
noun? It seems unlikely. Reading this sentence aloud reveals
its inherent problem: it is a *spoken* sentence, rambling and
informal, which might well be clear when it is given human
intonation. *They* (nod in the right direction) got too busy.
The word 'got' is unpleasant in sound and imprecise in
meaning, but we all use it in speech. Some guess-work is
needed in order to transform this sentence into a clear written statement, but we might feel that the following is as close
to the original meaning as possible:

Owing to pressure of work, faxes are no longer logged.

It is certainly a good deal shorter!
Good writing, then, avoids abstraction, and is direct and
uncomplicated. Since the written word lacks the stress given
by the human voice, it has to be clear, logical and precise. It
also demands variety. We have probably all read boring
passages in which sentences begin *First, we... Next,
we... Then... Then...* and so on. More than two sentences
which begin with the same words sound monotonous, as do
sentences which have very weak beginnings, such as *Another
example is when....* A little thought on the part of the writer
will usually produce a more interesting and encouraging
version of the sentence.
Readers need encouragement, especially that which comes
from seeing how a passage is constructed. 'Link' words and
phrases, which show a logical connection between sentences
or paragraphs, guide the reader through the document, and
incidentally help the flow of the writing. *At the same time, on
the other hand, meanwhile, bearing this in mind,* and, correctly used, *however,* introduce the next stage of the

argument. Some links add emphasis as well as flow; the earlier sentence about the engineers and the clients would read better with the addition of the small word *even*:

> *Even the engineers involved in the project were unsure of the client's identity.*

☐ 'Link' words and phrases guide the reader through the document and improve the flow of the writing.

This chapter will end with an example and a suggestion. A passage from the engineering student's work on progressive failure has been rewritten with the logical connecting links omitted. It is awkward to read, and the information is not easy to assimilate. This damaged version is then followed by the original version, and a comparison will show how much easier and more pleasant the latter is to read. Interestingly, this passage has not been written by a professional or even an experienced writer but by a student with a 'feel' for the flow of the language. This level of clear, fluent writing can be achieved by most engineers who want to produce a good style for everyday use. A suggestion for other engineering writers is that from time to time they read a passage of their writing out loud and listen, critically, to its sound. They may be surprised by obvious failings of style which can then be put right or perhaps, more pleasantly, by the easy, flowing style which they have achieved. 'Easy' style, incidentally, is easy only for the reader; it requires hard work from the writer!

☐ Listen critically to your own work.

Example of Passage without 'Links'

> *Structures which are highly optimised and operate at a high proportion of their ultimate load are most at risk from progressive failure. Obvious examples of this kind*

of structure are aerospace structures. Much attention has been paid in the design of airframes to ensuring fail-safety. They have to have as low a weight as possible and are optimised to ensure that no part is larger than necessary. A variation from the assumed load pattern could have a serious effect on members close to the site of damage. Local damage would have this effect. Allowance must be made for this eventuality. Fail-safe or damage-tolerant design will remain serviceable after having been damaged. Alternative load paths are incorporated into the structure by means of multiple redundancy. When damage occurs, there are a number of other members which can carry the extra load.

The Same Passage with 'Links'

Structures which are most at risk from progressive failure are those that are highly optimised and operate at a high proportion of their ultimate load. Aerospace structures are perhaps the most obvious example of such structures and it is in the design of air-frames that much attention has been paid to ensuring fail-safety. Since such structures must have as low a weight as possible, they are optimised to ensure that no part is larger than necessary. However, this means that a variation from the assumed load pattern, which would happen if local damage were sustained, could have a serious effect on members close to the site of damage if allowance is not made for this eventuality. Fail-safe or damage-tolerant design is that which will remain serviceable even after having been damaged. The main method by which this is achieved is the incorporation of alternative load paths into the structure by means of multiple redundancy. Thus when damage occurs, there are a number of other members which can carry the extra load.

Key Ideas

- Good style is appropriate to the needs of the reader (p. 104).

- Good style has its ABC: accuracy, brevity, clarity (p. 105).

- Use the active rather than the passive, whenever possible (p. 108).

- Follow national, international or professional conventions in presenting signs, symbols, units of measurement and abbreviations (p. 109).

- Diagrams should elucidate and not interrupt the message (p. 110).

- A well-chosen example helps the reader's understanding and gives life to the text (p. 111).

- References should show full and accurate bibliographical details, and should be consistent in form (p. 112).

- Avoid abstract words as far as possible. Use simple, direct language (p. 113).

- 'Link' words and phrases guide the reader through the document and improve the flow of the writing (p. 115).

- Listen critically to your own work (p. 115).

6 The Presentation of Written Information

The importance of good presentation ■ checking the facts ■ checking the text ■ the need for consistency ■ page layout ■ font size ■ line length ■ the use of space ■ title pages ■ binding ■ attractive and professional text

---

Earlier in this book (Chapter 4), we looked at the formats appropriate to memos, letters, reports and instructions, and at the conventions which should be followed in order to produce such documents with an appropriate structure.

There is another aspect, however, of the effectiveness of written information: its presentation. This is largely a matter of reader goodwill. If the document looks both professional and inviting, the reader will want to read it. If it looks scruffy and difficult to approach, the prospective reader will be put off, and will relegate the paper to the bottom of the pile, or, worse, to the wastepaper basket. A report which looks less than professional undermines the credibility of both writer and company; instructions which do not look 'official' may be ignored.

In this chapter, two different aspects of presentation are discussed: checking (to make sure that the correct information is correctly presented) and the layout of the printed page.

Most documents are checked for factual errors. In preparing to write, reference may be made to previous similar documents, to company guidelines, to books and journals in a library belonging to the company or an Institution.

Later, the writer goes back over the document, to make sure that its contents are technically correct and in accordance with company policy. A colleague's opinion may well be sought. In the case of a report or specification, a draft version will be passed up the hierarchy of the author's organisation so that it can be checked by more senior managers. In passing, it is worth mentioning that any changes made should always be for a good reason. Too many documents are subjected to a manager's need to 'make his/her mark' (the subconscious 'if I don't alter something, how can I show that I'm doing my job/how can I show how important my opinion is' response). Nevertheless, it is generally true that this stage is valuable in order to ensure that the document contains the right information logically presented, and that it is within the limits of what may be revealed; this is especially true when company confidentiality or security is involved.

In the case of instructions, the most sensible way to check what has been written is to ask a colleague to carry out the procedure under the writer's supervision; if a stage has been omitted or the wording is not clear, the problem should be immediately apparent. It is, of course, wise to choose a colleague who does not normally carry out such instructions, or the correct action might be taken automatically despite instructions to the contrary.

☐ All documents should be checked for factual accuracy.

The second stage of checking is often overlooked. The facts are assembled and the correct information has been obtained, but what appears on the page is not what the writer intended. Most engineers nowadays wordprocess their own material. They remain physically and mentally close to what they are producing; it is there on the computer on the desk, available for revision, additions or deletions at any time. The writer knows perfectly well what was intended, to the point where he or she will see it even if it isn't there.

We are all bad at checking our own work. Our interest and involvement is with the information, not with the wordprocessing, and it is all far too familiar to us. We fail to see what has gone wrong. There are two results of this failure, and both are serious for the individual and for the company in whose name the document is to be sent out. The first is that incorrect data are passed on to the reader (colleague, client, general public). It is easy to imagine the impact of a nought missed off the end of a price in an estimate or an invoice. It is impossible to guess at the amount of time and money wasted annually in telephone calls, meetings and discussions which take place to find out what is wrong with the product, system or payment, all because nobody bothered to check the original wordprocessing and to notice the mistake. It must run into millions of pounds. In the case of engineering information, the results of error may be even worse, in terms of industrial accidents and personal injury.

The second result of the failure to check what has been written is almost as serious, but even harder to quantify. It is the loss of professional credibility. The reader is, after all, taking for granted the expertise of the engineer who produced the document. Readers will order equipment on the basis of the information given; they will become involved with costly projects or stake the reputation of their companies or even their lives on the accuracy of the details they have been given. As soon as they see a mistake, however small and insignificant it may be in itself, they will start to worry about the accuracy of everything else. This is particularly true of mathematical information. If a letter in a word is misplaced or omitted, readers may well be able to guess what the word should be. The lasting effect of, for example, two letters transposed in a word, is to make readers doubt every *figure* in the writer's specification or quotation or test report. Everything is suspect and therefore potentially unreliable.

☐ Wordprocessing errors distort information and undermine credibility.
☐ All documents should be checked for accuracy of presentation.

There is no easy or complete answer to the problem of printing errors. Wordprocessors have built-in spellchecks, which are helpful provided that they are not used as the only form of checking. They will highlight nonsense words, but not the wrong word which makes sense or which is common throughout the document. *Now* instead of *not* is an outstanding example of the dangerous mistake which the spellcheck will ignore. It reverses the meaning and yet appears to make complete sense, as in:

> *The car is now safe to drive.*
> *The car is not safe to drive.*

This critical error will be picked up only by an alert human reader.

Ideally, every document should be checked specifically for errors. In practice, the level of checking will depend on the perceived importance of the document concerned. Memos and other comparatively casual writing may be checked quickly on the screen and as a printout, and this is probably sufficient. All reports and similar material which travel outside the company, and all specifications, instructions and manuals, should be thoroughly checked for printing errors as a separate activity from checking the facts. It is dangerous simply to combine these two operations, as inevitably the facts will appear to be more important. The best person to check the text itself is almost certainly a colleague who is familiar with the sort of information presented, without being an expert on the immediate subject discussed; it is important that this reader will not make assumptions about the writer's meaning. Diagrams, appendices and other such material must be included in the checking process, as should the title page, which is often overlooked.

Inconsistency has an effect similar to that of printing errors, in that it suggests a casual, slipshod approach to the writing. It is usually unimportant whether a writer chooses 's' or 'z' in a word like 'organisation', but a change of mind in the middle of the document undermines the reader's confidence. Even tiny changes, such as the more common *eg* rather than *e.g.* must be made consistently – and must be consistent also with one another, so that if *eg* is used, *etc* will not have a full stop either. Part of the process of checking is to watch for inconsistency, and, sensibly, to record decisions about style as they are taken. This is particularly important if the document is produced by a group of people, some of whom might, for example, use *subcontractors* rather than the slightly old-fashioned form *sub-contractors*. One member of the group should have editorial responsibility for making such decisions and notifying other writers early in the production of the document; checking will then be made easier.

□ Consistency suggests conscientiousness and reassures the reader.

The first stage of checking will always be the duty of the writer. If the text can be left alone for a couple of days, or preferably a week, the author is much more likely to see mistakes than if checking started only five minutes after the printout was produced. Checking is a time-consuming job (lack of time is the most common 'excuse' for not checking), and it is a mistake either to leave insufficient time to be thorough or to assume that a 'spare hour' can be devoted at the end to going through the complete document. Concentration is short, especially as checking is boring as well as time-consuming, and the writer should take regular short breaks. Even a couple of minutes spent every quarter of an hour in leaning back and looking out of the window will help. It is important to check not only from the screen but also from a printout – a much more 'normal' angle at which to read – and it is helpful to put a blank piece of paper over the text and to reveal one line at a time. It is unlikely that a single line will make complete sense, and so the writer is able to concentrate on the words rather than on the meaning. Some mistakes are particularly difficult to see, such as the duplication or omission of an 'unimportant' word like 'the', or the creation of a word which has the same shape as the word intended, such as 'casual' for 'causal' or 'form' for 'from.'

Wordprocessors are a wonderful invention, but they can produce their own problems. Just because it is so easy to correct a mistake, the writer whose inspiration is in full flow will tend to leave the correction until later, when the mistake may be much less obvious. As material is moved around, it may no longer be consistent with what now comes before it, so that a singular subject may end up with a plural verb, and so on. The insertion of an extra paragraph will move other material on different pages, and a heading may be separated from the text to which it refers, or a key word at the end of a sentence may now be on a separate page from its context. The task of checking includes making sure that such irritations have been removed.

Layout on the page can make both reading and checking easier or more difficult. Wordprocessors can produce a right-justified text, that is, a text in which all lines end at the same point, creating a regular right-hand margin. This may look attractive – though opinions vary – but it is harder to read and to check than unjustified ('ragged right') text. If numbers are included in a long paragraph, it is particularly important that the eye can move accurately from the end of one line to the beginning of the next, and is not allowed to jump a line.

☐ Right-hand justification should be avoided if possible.

Many journals divide the page into two or three columns. This creates a sense of immediacy and can look attractive, but it increases the number of awkward word-breaks, and makes even short paragraphs look too long. Engineers writing for professional journals should always study the layout of articles in the chosen journal before beginning to write. If the text appears in columns, the writer should shorten the paragraphs so that the text still looks easy to assimilate. Counting the number of words in previously published articles is a useful guide to the average number in a paragraph and on the printed page. (More advice about technical writing for publication is suggested in the Further Reading section at the end of this book.)

The font chosen should be large enough for ease of reading. On the whole, 10 or 11 point is a sensible size, with normal wordprocessor interline spacing; if mathematical material is included, special care should be taken with interline spacing to ensure that superscript and subscript figures do not corrupt the lines of text. In printing an early draft of the document, the writer may choose double spacing in order to have room to make notes at the appropriate point on the printout, but it will be a little harder to read in the finished product.

Line length should also be checked. Ideally, the number of key strokes to a line of print should be between 60 and 80; if there are more than this, the reader's eyes cannot move easily from the end of one line to the beginning of the next, and the text will be perceived as heavy to read, although the reader may well not know why this is so.

☐ Print size and line length should be chosen for the reader's convenience.

The need for space has been stressed several times in this book. The layout of the page should look well spaced, with appropriate gaps between words, lines, paragraphs or sections, and round the edges of the text. Margin space on the left is important so that words or figures are not lost in the binding, and there should be adequate space at the top and bottom of the page; a minimum of half an inch (just over 1 cm) is suggested for printed material and twice that amount for wordprocessed text. A congested page looks unattractive and heavy, and is unlikely to encourage the reader.

☐ Leave space on the page to allow the text to stand out clearly.

The title page is usually the first page of a document which the reader sees. Title pages give status as well as giving administrative information. Manuals, instructions, specifications and reports have title pages, which contain standard information: title, author and date. A reference may also be required, together with the issuing company's logo, name and address. Similar details of the client company for whom the document was produced will be included as appropriate. A statement of confidentiality should be shown if it is needed. All these details are often specified in the company's procedures, but the individual engineer may be able to ensure that the layout of the title page is uncluttered and that it gives a professional look to the whole document.

☐ A well laid-out title page gives a professional appearance to the whole document.

Many reports and similar documents will be photocopied, at least in part. At this stage, the value of numbering the pages is apparent – it is too easy for unnumbered pages to be omitted by accident. Poor quality copying can make figures unclear or ambiguous, and checking of one set of copied material is a good idea.

Binding is the last stage in the completion of most documents. The style of binding may be laid down by the company, but the choice should be made in the light of the document's importance, length and longevity. Stapling is inappropriate for all but the shortest in-company documentation: the last page is easily torn off, and the rule seems to apply that the staple itself will go through the most important word on the second page. A slide bar binding, usually chosen by students because it is cheap and can be re-used, has the disadvantage that pages will not lie flat on the desk, and the slide bar itself is likely to come off, allowing the pages to fall apart. Spiral bindings are widely used: they will lie flat on the table in use, they are secure and on the whole look attractive. They are not sensible for very short or very long reports, and if several documents are kept on a shelf together, the spirals can become entangled. Ring binders are more bulky but very useful for material which will have to be updated regularly; they have the associated disadvantage that it is all too easy for people to remove pages or sections without permission.

Other more permanent forms of binding tend to be reserved for major documents which have a long lifespan in printed form, or which are intended primarily to impress a client. Indeed, some documents would benefit from being professionally designed. If the information is to be used commercially, if it is very expensive to obtain or if the company's professional image is at stake, it is worth employing a designer so that the document makes the strongest possible impact on the recipient. Major published reports are clearly in this category.

☐ Both overall design and binding should be appropriate to the impact which the document is intended to make.

However the final document is produced, it should look and feel good in use. In the light of advances in technology, it may seem surprising that this is still so true; we might have been living by now with the 'paperless office' we expected, so that all documents are held only on computers. In practice, this has not yet happened. Documents are produced and updated by computer but generally transmitted in paper form, and it is likely to be some years before there is a widespread change. In the meantime, the impact of a document remains important. It should attract the attention of its readers and users by its attractive appearance, clear helpful format, accuracy of presentation, and precise, unambiguous written style. The impact it makes should always be one of pleasure and reassurance that time and trouble have been taken to produce a document worthy of its technical content.

☐ A document should look both attractive and professional; it should inspire confidence.

Key Ideas

■ All documents should be checked for factual accuracy (p. 119).

■ Wordprocessing errors distort information and undermine credibility (p. 120).

■ All documents should be checked for accuracy of presentation (p. 120).

■ Consistency suggests conscientiousness and reassures the reader (p. 122).

■ Right-hand justification should be avoided if possible (p. 123).

- Print size and line length should be chosen for the reader's convenience (p. 124).

- Leave space on the page to allow the text to stand out clearly (p. 124).

- A well laid-out title page gives a professional appearance to the whole document (p. 125).

- Both overall design and binding should be appropriate to the impact which the document is intended to make (p. 126).

- A document should look both attractive and professional; it should inspire confidence (p. 126).

Notes

1. Suchet, John, 'Further in his Cap', *Guardian*, 1 February 1988.
2. Hart, Geoff, 'The Five Ws: An Old Tool for the New Task of Audience Analysis', *Communicator* (Journal of the Institute of Scientific and Technical Communicators), n.s. vol. 5, no. 7 (Autumn 1996) pp. 12–17.
3. Maude, B., *Practical Communication for Managers* (Harlow: Longman, 1974) p. 65.
4. Radford, J. D., *The Engineer and Society* (London: Macmillan, 1984) p. 197.
5. van Emden, Joan and Easteal, Jennifer, *Technical Writing and Speaking: An Introduction* (Maidenhead: McGraw-Hill, 1996) pp. 38–42.
6. van Emden, Joan and Easteal, Jennifer, *Technical Report Writing* (Institution of Electrical Engineers Professional Brief), 4th edn (London: IEE, 1997) p. 4.

Further Reading

In spite of the computer spellcheck, an important tool for any writer is still a good dictionary, as up to date as possible (although all published dictionaries lag behind current usage, especially in the spoken language). Two sizes of dictionary are useful, one for the desk and one to be carried in a pocket or briefcase. Recommended are:

The Concise Oxford Dictionary
Longman Pocket English Dictionary
Collins Gem Dictionary of Spelling and Word Division (which has no definitions and therefore plenty of room for words).

A thesaurus aids the writer in the choice of words, as it presents the range of words available for a particular purpose; however, no thesaurus can show the slight shades of meaning which make us choose a particular word rather than its synonyms.

Roget's Thesaurus

is the standard work, while

Collins Gem Thesaurus

is small enough to be carried about.
   The best widely available book about punctuation, in spite of its age, is

G. V. Carey: *Mind the Stop*, Penguin, Harmondsworth, 1971.

For choice of words,

Ernest Gowers: *The Complete Plain Words*, Pelican, Harmondsworth, 1987

is widely used.
   Two longer books which look at writing style with the technical writer especially in mind are

Joan van Emden and Jennifer Easteal: *Technical Writing and Speaking: an Introduction*, McGraw-Hill, Maidenhead, 1996.

and

John Kirkman: *Good Style: Writing for Science and Technology*, Spon, London, 1992.

Editors of technical material will find the following of interest:

The Oxford Dictionary for Writers and Editors, Oxford University Press, Oxford, new edition to be published 1998.

Engineers will themselves know of technical dictionaries which apply to their own specialisms. Two such are

John S. Scott: *Dictionary of Civil Engineering*, Penguin Reference Books, Harmondsworth, 1991.

IEEE Standard Dictionary of Electrical and Electronics Terms, IEEE, New York, 1993.

The *BSI Glossaries* are of general use for engineering terms, as is

Chambers Materials Science and Technology Dictionary, Chambers, Edinburgh, 1993.

The Engineering Institutions offer a wide range of advice to their members, and their librarians will give particular help, if asked. The Institution of Electrical Engineers publishes a series of Professional Briefs, of which the most relevant are:

Joan van Emden and Jennifer Easteal: *Technical Report Writing*, 4th edn, 1997.
Technical Writing for Publication, 1996.
Units and Symbols for Electrical and Electronic Engineering, 1992.

These can be obtained through the Courses Unit of the IEE at Stevenage (telephone: 01438 767288/9, fax: 01438 742856), which also holds regular courses on technical report writing.